U.S. ARMY
COUNTERINTELLIGENCE
HANDBOOK

U.S. ARMY COUNTERINTELLIGENCE HANDBOOK

DEPARTMENT OF THE ARMY

SKYHORSE PUBLISHING

All inquiries should be addressed to Skyhorse Publishing, 307 West 36th Street, 11th Floor, New York, NY 10018.

Skyhorse Publishing books may be purchased in bulk at special discounts for sales promotion, corporate gifts, fund-raising, or educational purposes. Special editions can also be created to specifications. For details, contact the Special Sales Department, Skyhorse Publishing, 307 West 36th Street, 11th Floor, New York, NY 10018 or info@sky-horsepublishing.com.

Visit our website at www.skyhorsepublishing.com.

10 9 8 7 6 5

Library of Congress Cataloging-in-Publication Data is available on file.

ISBN: 978-1-62087-478-3

Printed in the United States of America

Contents—FM 34-60

Preface

This field manual (FM) provides guidance to commanders, counterintelligence (CI) agents, and analysts. The first four chapters provide information to the commander and staff while the remainder provides the tactics, techniques, and procedures (TTP) required to aggressively identify, neutralize, and exploit foreign intelligence attempts to conduct operations against the United States (US) Army.

CI supports Army operations by providing a clear picture of the threat to commands at all echelons and actions required to protect the force from exploitation by foreign intelligence. CI operations include conducting investigations, offensive and defensive operations, security and vulnerability analyses, and intelligence collection in peace and at all levels of conflict to support command needs.

CI supports the total intelligence process by focusing on foreign intelligence collection efforts. CI is designed to provide commanders the enemy intelligence collection situation and targeting information in order to counter foreign intelligence service (FIS) activities. CI is an integral part of the US Army's all-source intelligence capability.

By its nature, CI is a multidiscipline effort that includes counter-human intelligence (C-HUMINT), counter-signals intelligence (C-SIGINT), and counter-imagery intelligence (C-IMINT) designed to counter foreign all-source collection. The CI force in conjunction with other intelligence assets must have the capability to detect all aspects of intelligence collection and related activities that pose a threat to the security of Army operations, personnel, and materiel.

Through its database (friendly and adversary) and analytical capability, CI provides sound recommendations, which if implemented, will result in the denial of information to the threat.

It should be noted that any decision regarding the implementation of CI recommendations aimed at denying collection opportunities to the adversary is a command decision. The commander may decide to accept the risk of enemy collection in favor of time, resources, or other higher priority considerations. At that point, the CI mission is considered to be successful because it is a tool of the commander.

This manual is designed for use by commanders and their staffs; all military intelligence (MI) commanders, their staffs, and trainers; and MI personnel at all echelons. It applies equally to the Active Army, United States Army Reserve (USAR), and Army National Guard (ARNG). It is also intended for commanders and staffs of joint and combined commands, United States Naval and Marine Forces, units of the US Air Force, and the military forces of allied countries.

Provisions of this manual are subject to international Standardization Agreements (STANAGs) 2363 and 2844 (Edition Two). When amendment, revision, or cancellation of this publication affects or violates the international agreements concerned, the preparing agency will take appropriate reconciliation action through international standardization channels. Chapter 1 implements STANAG 2844 (Edition Two) and Chapter 3 implements STANAG 2363.

The proponent of this publication is the United States Army Intelligence Center and Fort Huachuca. Send comments and recommendations on DA Form 2028 (Recommended Changes to Publications and Blank Forms) directly to Commander, US Army Intelligence Center and Fort Huachuca, ATTN: ATZS-TDL-D, Fort Huachuca, AZ 85613-6000.

Unless this publication states otherwise, masculine nouns and pronouns do not refer exclusively to men.

Chapter 1

Mission and Structure

General

Threat intelligence services have the capability to conduct continuous collection against the US Army during peacetime, operations other than war (OOTW), and during war itself. The intelligence that results from these operations provides a significant advantage to threat forces, and could easily result in increased US casualties on the battlefield. Fortunately, there are many actions we can take to counter threat intelligence efforts and to provide force protection to all US Army units. The most dramatic of these actions are designed to neutralize enemy collection. These actions include

■ Using field artillery to destroy ground-based enemy signals intelligence (SIGINT) collectors.

■ Conducting sophisticated C-HUMINT operations in a foreign city long before overt hostilities commence.

■ Employing direct fire weapon systems to destroy enemy reconnaissance. Brigades conducting defensive operations at the National Training Center often commit a tank-infantry company team to provide counterreconnaissance, intelligence, surveillance, and target acquisition (C-RISTA) protection.

While not as flashy, routine security procedures provide crucial force protection. These procedures include but are not limited to

■ Personnel security, to include background investigations, will ensure all personnel who have access to sensitive or classified information will fully protect it.

■ Information security, particularly in regard to handling classified and compartmented information, will be a challenging field in the future considering the ease with

1

which information can be copied and transmitted in an increasingly automated Army.

- Physical security, which ensures physical measures are taken to safeguard personnel, prevents unauthorized access to equipment, installations, materiel, and documents to safeguard them against espionage, sabotage, damage, and theft.

- Operations security (OPSEC), which ensures that all essential elements of friendly information (EEFI), are reasonably concealed from enemy collection assets.

Another crucial component in the fight against threat collection efforts is CI analysis. These include efforts to identify the general capabilities and specific operations of enemy human intelligence (HUMINT), SIGINT, and imagery intelligence (IMINT) collection. CI analysis also includes the development of profiles that identify friendly vulnerabilities to enemy collection and possible countermeasures.

Measures such as these provide a crucial force protection shield that is difficult for the FIS to penetrate. More importantly, a comprehensive CI program significantly degrades the threat's ability to target and conduct combat or terrorist operations against US Forces. Total CI provides the combat commander with a definite advantage on the battlefield.

AR 381-10, AR 381-12, and AR 381-47 (S) contain policies and procedures governing the conduct of intelligence activities by Department of the Army (DA).

Mission

The CI mission is authorized by Executive Order (EO)12333, implemented by AR 381-20. The Army conducts aggressive, comprehensive, and coordinated CI activities worldwide. The purpose is to detect, identify, assess, counter, neutralize, or exploit threat intelligence collection efforts. This mission is accomplished during peacetime and all levels of conflict. Many CI functions, shown in Figure 1-1, are conducted by echelons above corps (EAC); some by echelons corps and below (ECB); and some are conducted by both. Those CI assets found at ECB respond to tactical commanders. EAC assets respond primarily to commanders of intelligence units

FUNCTION	EAC			ECB		
	PEACE	WAR	OOTW	PEACE	WAR	OOTW
INVESTIGATIONS						
Personnel Security (OCONUS)	X	X	X	X	X	X
Army CI Investigations:						
Treason	X	X	X		X	
Espionage	X	X	X	X	X	X
Spying		X	X			
Subversion		X				
Sedition		X				
FIS-directed sabotage	X	X	X	X	X	X
Terrorism	X	X	X	X	X	X
Assassination	X	X	X	X	X	X
Detection	X	X	X	X	X	X
Detention	X	X	X	X	X	X
Special category absentees	X	X	X	X	X	X
Deliberate security violations	X	X	X	X	X	X
Suicide or attempted suicide	X	X	X	X	X	X
CI scope polygraph examinations	X	X	X			
Technical penetration	X	X	X			
OPERATIONS						
CI special operations (AR 381-47 (S))	X	X	X			
CI support to force protection:	X	X	X	X	X	X
CI support to mobilization	X	X	X	X	X	X
CI support to combatting terrorism	X	X	X	X	X	X
CI support to rear operations		X	X		X	X
CI support to civil-military operations		X	X		X	X
CI support to psychological operations		X	X		X	X
CI support to battlefield deception		X	X		X	X
CI support to OPSEC	X	X	X	X	X	X
CI support to friendly C-E	X	X	X	X	X	X
CI support to information operations	X	X	X	X	X	X
CI support to counter-drugs	X		X	X		X
CI force protection source operations (deployed)	X	X	X	X	X	X
Advice and assistance	X	X	X	X	X	X
CI technical support activities	X	X	X			
CI support to acquisition and SAPs	X	X	X			
CI support to HUMINT	X	X	X			
CI support to treaty verification	X	X	X	X		X
Liaison	X	X	X	X	X	X
CI support to domestic civil disturbance			X			X
CI support to natural disaster operations			X			X
C-SIGINT support	X	X	X	X	X	X
C-IMINT	X	X	X	X	X	X
Hostile intelligence simulation (Red Team)	X	X	X			
Covering agent support	X	X	X	X	X	X
COLLECTION						
Identifying and validating requirements	X	X	X	X	X	X
Local operational data collection	X	X	X	X	X	X
Debriefing and interrogation	X	X	X	X	X	X
Returned US defector debriefing	X	X	X	X		X
ANALYSIS AND SYNTHESIS						
Threat and friendly databases	X	X	X	X	X	X
Threat assessment	X	X	X	X	X	X
Vulnerability assessment	X	X	X	X	X	X
Countermeasures recommendations	X	X	X	X	X	X
Countermeasures evaluation	X	X	X	X	X	X

Figure 1-1. Counterintelligence functions.

while supporting all commanders within their theater or area of operations (AO).

The essence of the Army's CI mission is to support force protection. By its nature, CI is a multidiscipline (C-HUMINT, C-SIGINT, and C-IMINT) function designed to degrade threat intelligence and targeting capabilities. Multidiscipline counterintelligence (MDCI) is an integral and equal part of intelligence and electronic warfare (IEW). MDCI operations support force protection through OPSEC, deception, and rear area operations across the range of military operations. For more information on IEW operations, see FM 34-1.

CI in Support of Force XXI

CI must meet the goals and objectives of Force XXI and force projection operations. US Forces will be continental United States (CONUS)-based with a limited forward presence. The Army must be capable of rapidly deploying anywhere in the world; operating in a joint or combined (multinational) environment; and defeating simultaneous regional threats on the battlefield; or conducting OOTW. CI, as part of IEW, is fundamental to effective planning, security, and execution of force projection operations. Successful force projection CI support is based on the same five key principles shown in Figure 1-2 and discussed below. CI, in support of force protection, will be required on the initial deployment of any force projection operation.

The Commander Drives Intelligence:

The commander focuses on the intelligence system by clearly designating his priority intelligence requirements (PIR), targeting requirements and priorities. He ensures that the Intelligence Battlefield Operating System (BOS) is fully employed and synchronized with his maneuver and fire support BOSs. He demands that the Intelligence BOS provides the intelligence he needs, when he needs it, and in the form he needs.

Intelligence Synchronization:

The J2 or G2 synchronizes intelligence collection, analysis, and dissemination with operations to ensure the commander receives the intelligence he needs, in the form he can use, and in time to influence the decisionmaking process. Intelligence synchronization is a continuous process which keeps IEW operations tied to the commander's critical decisions and con-cept of operations. CI collection, analysis, and dissemination, like other intelligence, have to meet the commander's time requirements to be of any use other than historical.

Split-Based Operations:

Split-based operations provide deploying tactical command-ers with a portion of their collection assets and augment full employment of organicassets. Split-based intelligence operations employ collection and analysis elements from all echelons, national to tactical, in sanctuaries from which they can operate against the target area.

Tactical Tailoring:

In force projection operations, the commander tactically tailors CI, as well as all IEW, support for each contingency based on the mission and availability of resources. He must decide which key CI personnel and equipment to deploy early, and when to phase in his remaining CI assets.

Broadcast Dissemination:

Broadcast dissemination of intelligence includes the simulta-neous broadcast of near-real time (NRT) CI from collectors and processors at all echelons. It permits commanders at all echelons to simultaneously receive the same intelligence, thereby providing a common picture of the battlefield. It allows commanders to skip echelons and pull CI directly from the echelon broadcasting it.

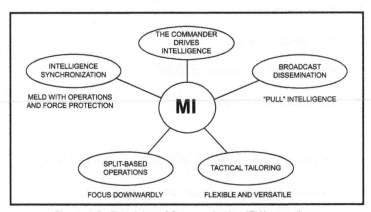

Figure 1-2. Principles of force projection IEW operations.

5

Intelligence Tasks

Army military intelligence (MI) accomplishes its mission by performing six primary tasks: providing indications and warnings (I&W); performing intelligence preparation of the battlefield (IPB); performing situation development; supporting target development and targeting; developing force protection intelligence; and performing battle damage assessment (BDA).

CI Tasks

The role of CI is to support the commander's requirements to preserve essential secrecy and to protect the force directly or indirectly. Thus, CI contributes to the commander's force protection programs. Force protection is a command responsibility to protect personnel, equipment, and facilities. To carry out his force protection responsibilities, a commander requires support from several sources, one of which is the intelligence community. CI support to force protection must be tailored to the sensitivity of the supported organization and its vulnerability to FIS and hostile attack. CI support can be tailored from a combination of activities to include

- Mobilization security, including ports and major records repositories.

- Combatting terrorism.

- Rear operations.

- Civil-military affairs.

- Psychological operations (PSYOP).

- Battlefield deception.

- OPSEC.

- Friendly Communications-Electronics (C-E) (C-SIGINT).

- CI force protection source operations (CFSO).

6

Army CI as a Function of MI

Army CI, as a multidiscipline intelligence function, is an integral part of the Army and Department of Defense (DOD) and national intelligence communities. CI missions are conducted in support of the objectives of these communities.

Counterreconnaissance

CI is an integral part of the command counterreconnaissance effort. Human and other intelligence sensors determine adversary reconnaissance, intelligence, surveillance, and target acquisition (RISTA) and other battlefield capabilities, and project resultant data into battle planning and execution. As the adversary worries about our C-RISTA capability, our CI efforts target his RISTA capabilities. CI focuses on the HUMINT threat in the AO and provides analytical support in identifying enemy SIGINT and IMINT capabilities and intentions. CI has a limited neutralization and exploitation capability directed at low-level adversary HUMINT collectors or sympathizers acting in a collection or sabotage capacity. The commander is responsible for security countermeasure programs and training to include personnel, physical, document, information security, crime prevention, and OPSEC.

Other Specialties

Army CI is not limited to the activities of a small force of CI agents and technicians; rather, it is the responsibility of all Army personnel to follow common sense security measures to minimize any foreign intelligence threat. Although a major part of the CI mission is to counter or neutralize FIS efforts, this does not mean that only CI personnel take part in these actions. They may require

■   Other intelligence specialists such as interrogators.

■   Military police (MP).

■   Civilian counterparts and authorities.

■   Combat forces.

■   Civil-military affairs and PSYOP.

The combined use of C-HUMINT, C-SIGINT, and C-IMINT TTPs provides a multidisciplined approach to denying information to unauthorized persons. This approach limits the threat's ability to collect against us. Although this FM describes these three operations separately in Chapter 3, they are often conducted simultaneously by the same assets.

Peace, War, and OOTW

The Army conducts CI during peacetime and at all levels of conflict to protect the force from foreign exploitation. During peacetime, CI simultaneously supports the commander's needs and DA policy.

During war, CI operations are much the same as in peacetime, except the adversary state or nation is well-defined. The commander's needs are the top priority.

OOTW may include the direct or indirect support of one or more foreign governments or groups, or international organizations such as the North Atlantic Treaty Organization (NATO). OOTW may be initiated unilaterally in the absence of foreign support. Whether unilateral or multinational, US Forces usually operate in a joint environment. Normally in OOTW, military force is used only as a last resort. OOTW consists of the following operational categories:

- Noncombatant evacuation operations.

- Arms control.

- Support to domestic civil authorities.

- Humanitarian assistance (HA) and disaster relief.

- Security assistance.

- Nation assistance.

- Support to counter-drug operations.

- Combatting terrorism.

- Peacekeeping operations.

- Peace enforcement.

- Show of force.

- Support for insurgencies and counterinsurgencies.

- Attacks and raids.

The CI Structure

To accomplish the CI mission at various echelons, specially trained CI personnel are assigned to tactical CI organizations as shown in Figure 1-3. Organizations include

- CI organizations organic to theater Army MI brigades or groups which are United States Army Intelligence and Security Command (INSCOM) organizations.

- Tactical exploitation battalion (TEB) and headquarters (HQ) and operations battalion of the corps MI brigade.

- MI battalion at division.

- MI companies at armored cavalry regiments (ACRs) and separate brigades.

- MI elements at special forces groups.

At each echelon, CI teams provide command and control (C 2) of CI assets; conduct CI investigations, operations, and collection; perform analysis and produce CI products; and provide security advice and assistance.

Only CI officers, technicians, agents, or accredited civilian employees control and conduct investigations. Additionally, DA policy identifies CFSO as a CI function as described in Chapter 4. CI personnel are also collectors of information, working individually or in teams with interrogators and technicians when resources permit. At ECB, CI personnel work in CI platoons at division level and CI companies at corps level. At EAC, CI personnel work individually or in groups in field offices, resident offices, or MI detachments or companies.

Another CI military occupational specialty (MOS) is MDCI analyst 97G. In addition to performing C-SIGINT operations

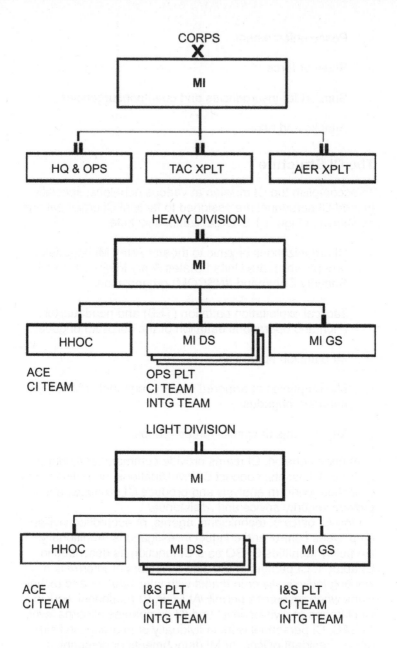

Figure 1-3. CI organization at echelons corps and below.

and communications monitoring, these soldiers perform MDCI analysis and produce MDCI products.

Interrogator and other personnel possessing requisite linguistics capability work with CI teams when conditions and resources permit. Provided these personnel are competent in the foreign language appropriate to the AO and possess the required security clearance, they perform interpreter, translator, liaison officer, and (after appropriate training) source handler duties. CI teams, found in some tactical units, are task organized based on mission, enemy, troops, terrain and weather, and time available (METT-T) factors. CI teams are composed of a CI technician, several CI agents and interrogation personnel, and MDCI analysts. Other CI teams are composed of CI agents and analyst personnel. At EAC, CI personnel work individually or in groups in resident offices, field offices, MI detachments, companies, or regions. Depending on the mission, additional specialists may either be attached or task organized to provide temporary expertise, such as

■ Physical security specialists.

■ MP.

■ Other intelligence personnel trained to accomplish a specific mission.

CI Support to US Forces

CI assets are deployed to provide area coverage. Only when driven by PIR should they be given a mission other than area coverage, such as specialized support to a special access program. When assigning missions to CI elements, METT-T must be carefully considered to ensure tasks are prioritized and CI assets are properly utilized.

Most CI operations develop slowly. Therefore, missions should be assigned for relatively long periods. If a team is investigating a sabotage incident, its mission should be assigned for as long as it takes to complete the mission. If a team is establishing liaison with host nation officials, this mission should remain with the team long enough to turn the liaison over to another team.

11

Within corps and divisions, CI assets are given an area coverage role. Based on priorities established by the corps or division commander, or G2, the MI battalion commander controls the CI assets as they execute the mission.

Although CI operations may change with priorities, CI assets must attempt to ensure commanders get what they need, when they need it, and in a form they can use before changing missions.

Planning

For contingency operations, CI elements should have the following procedures firmly in place prior to deployment:

■ Updated threat databases.

■ Planned CI communications in time to support decisionmaking.

■ Approved operations plans (OPLANs) with financial annexes for any source operations; for example, CFSO and host country liaison.

■ Appropriate and up-to-date country studies.

■ Established intelligence contingency funds (ICF). See AR 381-141.

■ Ongoing contact with theater CI elements to facilitate exchanging information and, where applicable, passing assets after employment.

■ Team reaction time must be rapid since contingencies can occur in locations with no US presence and with little warning. Therefore, the team needs a good working relationship with elements maintaining CI databases. The team should have a generic plan or established standing operating procedures (SOPs) which vary according to the type of OOTW. Teams would not do exactly the same things in peacekeeping, peace enforcement, and CI support to treaty verification, domestic civil disturbances, and natural disaster operations. Generally, the team must consider what is appropriate and feasible.

■ Procedures which tailor intelligence support packages to support planning and contingency operations should refer to the National Ground Intelligence Center (NGIC) and the 902d MI Group CI Analysis Center (CIAC) as a source of data.

Tasking and Reporting

CI teams receive taskings based on requirements from higher echelon. Taskings are normally generated by collection managers based on command needs or information gaps in analytical holdings and provided to the commander of the CI assets. CI teams also request information from organizations (such as HUMINT, SIGINT, or IMINT collectors) in support of CI missions.

Units report information based on the mission assigned and information collected. Information is reported to the CI unit's parent unit for C 2. It is also reported to others as directed by the parent unit, depending on the situation, mission, and the type and sensitivity of information collected. Other recipients of information could include

■ Theater J2 or Joint Task Force CI Coordinating Authority (JTFCICA), if applicable.

■ G2 or Deputy Chief of Staff, Intelligence (DCSINT) of theater ground component commander.

■ Other units, as appropriate, approved by the MI unit commander.

Joint and Combined Operations

Any future conflict may involve joint and combined operations. See FM 34-1. While operational control of intelligence elements normally rests with component commanders, joint staffs coordinate and provide operational guidance. In the case of CI operations, the J2, through the JTFCICA in accordance with Joint Publication 2-01.2, establishes areas of responsibility for CI operations for component forces;

produces and disseminates CI products by integrating information provided by subordinate commands; and coordinates and obtains intelligence and CI support from national
agencies and disseminates this information to operational
commands.

Joint Operations:

Most contingency operations require the deployment of joint
forces. Deployment is directed by the national command
authority. When this occurs, CI elements from various services (component commands of the joint command) may be
included in the task organization. The joint force or task force
commander and staff must either identify specific operational
boundaries or combine CI assets under a common command. This ensures continuity of effort without duplication.

If the IEW organization does not include CI assets and CI
assets are added from nonorganic organizations, they should
be assigned under the operational control of the commander
of the IEW organization. This clearly identifies C 2 relationships. For more information on IEW organization, see Joint
Publication 2-01.2.

Combined Operations:

The US Army and its allies conduct combined CI operations to attain mutual objectives. In many cases they conduct
these operations because the operating area does not permit
Americans to conduct unilateral operations.

Combined operations are required by Status of Forces
Agreements (SOFAs). In either case, the objective is the
same—to counter the intelligence capabilities and operations
of our adversaries. Combined CI operations include

- Investigations.

- CI support to deception.

- CI support to OPSEC.

- CFSO.

- Security advice and assistance.

- CI analysis and production.

Combined CI operations range from a mere exchange of information to conducting integrated CI operations. Local policy and procedures dictate the extent to which combined operations are conducted.

Special Operations:

CI is a critical component of intelligence support to Army Special Operations Forces (ARSOF). There are CI and inter-rogation assets included in the organic structure of ARSOF units. CI elements conduct tactical HUMINT collection, analy-sis and production, and other operations support activities which include but are not limited to

■ CI support to OPSEC.

■ CI support to deception.

■ CI liaison.

■ CFSO.

■ Limited investigations.

CI support to ARSOF is conducted throughout the phases of force protection operations during peace, war, and OOTW and must be tailored. For more information on CI operations, see FM 3-45 (S).

Legal Review

CI activities are authorized by law and regulation. Commanders and CI personnel should coordinate CI activities with the servicing judge advocate for verification of compliance with law and regulation prior to execution. Where appropriate, judge advocates should be part of the planning process.

Chapter 2

Investigations

General

During the last decade, a record number of espionage cases have involved US service members and US Government civilian employees. Most of these cases involved "volunteers," individuals who sought FIS employment. The US Army's authority to conduct intelligence and CI activities is EO 12333. Implementation of the EO and applicable DOD directives by the US Army are in AR 381-20. To accomplish this mission, CI agents conduct investigations to enhance, maintain, and preserve the security of the US Army. See Appendix A for TTPs.

Types of Investigations

US Army CI forces conduct two types of investigations: CI investigations (also known as Subversion and Espionage Directed Against US Army and Deliberate Security Violations [SAEDA] investigations) and personnel security investigations (PSIs).

CI Investigations:
CI investigations are national security investigations under Army CI jurisdiction. AR 381-12 requires US Army personnel, both civilian and military, to report suspicious activities. This program provides the greatest number of leads to Army CI personnel. AR 381-20 specifically lists those activities which fall under Army CI investigative jurisdiction. Types of CI investigations include

- Treason.

- Espionage.

- Spying.

- Subversion.

- Sedition.

- FIS-directed sabotage.

- CI aspects of terrorist activities directed against the Army.

- CI aspects of assassination or incapacitation of Army personnel by terrorists or by agents of a foreign power.

- Defection of military and DA civilians overseas, and debriefing of the individual upon return to US control.

- Detention of Army military or civilian personnel by a government or hostile force with interests contrary to those of the US.

- Investigation and debriefing of military members and Army civilians overseas who are declared absent without leave (AWOL), missing, or deserters who had access to TOP SECRET national defense information or sensitive compartmented information (SCI) within one year prior to their being declared AWOL, missing, or deserters. Investigation and debriefing of special category absentees (SCA) who were in a special mission unit (SMU); who had access to one or more special access programs (SAPs); or were in the DA Cryptographic Access Program (DACAP).

- CI aspects of security violations and compromises, and communications security (COMSEC) insecurities.

- CI aspects of incidents of DA personnel with a SECRET or higher security clearance, access to a SAP or SCI, or in the DACAP or SMU who commits or attempts to commit suicide.

- CI aspects of unofficial travel to designated countries, or contacts with foreign diplomatic facilities or official representatives, by all military personnel or DA civilians employed overseas.

- CI scope polygraph examinations. See AR 381-20.

- CI technical surveillance countermeasures (TSCM). See AR 381-14 (S).

All CI investigations are controlled by the Army Central Control Office (ACCO) and network of worldwide sub-control offices (SCOs). The ACCO and SCOs were established and are operated by Headquarters, INSCOM, at the direction of the DCSINT, Headquarters, DA.

ACCO. The ACCO is located at Fort Meade, MD. The ACCO exercises technical control, review, coordination, and oversight of Army CI controlled activities, including investigations. The ACCO has the specific responsibility to

■ Provide objective technical review ensuring complete and proper accounting of CI controlled activities in compliance with established policy.

■ Open CI investigations and assign all case control numbers; counterespionage (CE) projects; and ensure all initial SAEDA reports are reviewed in accordance with AR 381-47 (S).

■ Provide CFSO.

■ Retain the authority to assume direct control of any investigation or reassign it to another SCO. Direct case control includes imparting operational or investigative guidance, direct tasking, and investigative plan approval.

■ Establish a suspense system for timely completion of priority investigations.

■ Ensure that information within the purview of other intelligence, security, or law enforcement agencies is properly referred to them, and accept and process referrals from other agencies.

■ Ensure that case files and other records and reports are properly processed and transferred to the US Army Investigative Records Repository (IRR). Close all investigations upon files transfer to the IRR.

■ Provide quality control of reports to the SCOs.

■ Monitor status of suspended or transferred investigations in which Army interest continues.

■ Approve or disapprove requests for CI case summaries and declassification of classified case summaries, and declassify information classified by AR 381-47 (S).

Coordinate with other service CI agencies and the Federal Bureau of Investigation (FBI) when non-Army case summaries are requested.

- Conduct additional activities as stated in AR 381-47 (S).

SCOs. The SCOs are responsible for the day-to-day management of CI controlled activities. The SCO will

- Serve as the central focal point for monitoring the conduct of CI controlled activities within an area of responsibility.

- Open and close investigations, and determine their direction and scope, except those directly controlled by the ACCO. Approve investigative plans and task investigative elements accordingly.

- Respond to and coordinate technical direction and tasking from the ACCO.

- Report immediately to the ACCO all incidents which meet the criteria in AR 381-12 or AR 381-47 (S).

- Ensure all reports are properly addressed and disseminated based on content.

- Review all CI reporting to ensure it is accurate, complete, and in compliance with CI policy and intelligence oversight.

- Coordinate activities with legal and intelligence oversight officials, as required.

- Refer information within the purview of other intelligence, security, or law enforcement agencies in the AOR, as required.

- Coordinate briefings to commanders and senior intelligence officials.

- Pass lateral leads to other SCOs, when required, with information copies to the ACCO.

■ Complete and terminate activities and transmit final reports to the ACCO for case closure. Provide ACCO-approved summaries of information, when required, to the concerned commander and other agencies in the theater.

■ Forward requests for CI case summaries, non-Army case summaries, and requests for declassification to the ACCO.

■ Ensure appropriate coordination with (keep the respective senior intelligence officer informed) CI controlled activities affecting theater security.

■ Conduct additional activities as stated in AR 381-47 (S).

CI investigations are conducted only by CI personnel assigned or attached to units with a CI investigative mission. They have been school-trained by the United States Army Intelligence Center and Fort Huachuca (USAIC&FH) and hold specialty skill identifier 35E, or MOS 351B or 97B, and have been issued badge and credentials; or by Army civilian employees in career field 0132 who are assigned to CI units, have been school-trained, and are issued badge and credentials. See AR 381-20, paragraph 9-4.

Local national investigators employed by overseas Army CI units who have been issued MI representative credentials may conduct investigative leads. They will not be the primary or sole investigator.

Personnel Security Investigations:

AR 380-5 states that no person is entitled to knowledge of classified defense information or material solely by virtue of grade or position. This knowledge is entrusted only to those individuals whose official duties require access to the information. Persons requiring access must be investigated and receive favorable results before being cleared to receive classified material or information. Some foreign nationals and indigenous employees in overseas areas require a degree of access to perform their duties. Policy prohibits granting security clearances but permits granting of limited access authorization (LAA) following a favorable background investigation. These investigations are called LAA investigations and usually involve use of the polygraph.

PSIs are conducted outside continental United States (OCONUS) by the US Army on behalf of the Defense Investigative Service (DIS). Agents conduct PSIs to determine loyalty to the US and suitability for access to classified defense information. The investigative activity for PSIs runs the gamut from records checks, to interviews with people who know the subject of the investigation, to interviews with the subjects themselves. The complete guide for conducting PSIs is in DIS Manual 20-1-M.

Chapter 3

Operations and Techniques

General

Counterintelligence is information gathered and activities conducted to protect against espionage, other intelligence activities, sabotage, or assassinations conducted by or on behalf of foreign governments or elements of foreign organizations, persons, or international terrorist activities. See Joint Publication 1-02. This chapter identifies and focuses on CI operations and techniques as they support force protection, operations, and other military and civilian requirements.

Operations

There are two types of CI operations: *special operations and general operations.* Special operations involve direct or indirect engagement with FIS through human source or technical efforts. General operations are usually defensive in nature and are aimed at supporting force protection programs and formal security programs of Army commanders at all levels.

Essentially, all CI operations and activities support force protection. CI operations are not operations in the conventional combat arms sense. CI activities that do not fall under the other functional areas of investigations, collection, or analysis and production are characterized as operations. CI personnel are soldiers first. They are not equipped or trained to conduct standard military operations as a unit nor replace combat arms units, other combat support, or combat service support personnel. CI personnel support operations in peace, war, and OOTW. It is the commander's responsibility to direct execution. Once the decision to execute is made, operations will generally be carried out by combat forces. For example, in conflict, CI may identify threat collection assets that are legitimate tactical targets and recommend neutralization by appropriate artillery or air defense artillery fires.

CI information is developed through the intelligence cycle. The cycle consists of five phases: planning and directing, collecting, processing, producing, and disseminating. It is a continuous process and, even though each phase is conducted in sequence, all phases are conducted concurrently.

CI information without proper dissemination and action is useless. It is the role of the MDCI analyst working with other CI and intelligence specialists in the analysis and control element (ACE) to analyze threat data and determine counter-measures to deny the enemy collection capabilities or other threats. CI personnel recommend countermeasures through the S2/G2/J2 to the commander.

Special Operations:

CI special operations are generally carried out under the aus-pices of the National Foreign Counterintelligence Program. Normally, these operations involve direct or indirect engage-ment of FIS through human source or technical means. CI special operations are governed by AR 381-47(S) and consist of offensive CI operations, CE projects, and defensive source programs. Normally, these operations involve engagement of a FIS. Only those agencies approved by Headquarters, Department of the Army (HQDA), in implementing classi-fied memorandums conduct defensive source programs. Defensive source programs are designed to protect desig-nated Army activities against confirmed HUMINT threat.

General Operations:

As stated in Chapter 2, CI general operations are essen-tially defensive in nature and are aimed at supporting the force protection programs and formal security programs of Army commanders at all levels. Included in general operations are

- Advice and assistance programs.

- Technical support activities.

- Support to acquisition and SAP.

- Support to intelligence disciplines.

- Support to treaty verification.

■ Support to domestic civil disturbances.

■ Support to natural disaster operations.

■ Support to HUMINT.

Advice and Assistance Programs. Advice and assistance programs are conducted by CI teams at all levels to improve the security posture of supported organizations. These programs aid security managers in developing or improving security plans and SOPs. This support can be programmed or unprogrammed. Advice and assistance can help identify and neutralize threats to security from FIS or others who attempt to obtain information about US Army forces, programs, and operations. They provide threat information and identify specific vulnerabilities to security beyond the capability of a security manager. Advice and assistance can include but is not limited to

■ Conduct of inspections, security planning, the resolution of security problems, or development of classification guides.

■ CI surveys, technical inspections, and preconstruction technical assistance.

■ SAEDA training, providing SAEDA materials, and training security managers in the SAEDA programs.

■ Security investigations under AR 15-6 and AR 380-5.

Technical Support Activities. Technical support activities include TSCM, TEMPEST, polygraph, counter-surreptitious entry, and C-SIGINT (COMSEC monitoring). TSCM are specialized CI investigations governed by AR 381-14 (S). Intelligence polygraph is a technical investigative technique or tool and is governed by AR 195-6 and AR 381-20. TSCM and polygraph apply to intelligence as well as CI operations. INSCOM and the 650th MI Group conduct TSCM.

INSCOM conducts intelligence polygraph. For more information on polygraph, see Section III to Appendix A.

Automated Systems Security Automated systems security includes all of the technological safeguards and managerial

25

procedures established and applied to computer hardware, software, and developed data. INSCOM, under the technical direction of DCSINT (DAMI-CI), operates the automatic data processing system security enhancement program (ADPSSEP). Through that program, INSCOM has evaluation teams available to visit Army, and as directed, selected DOD contractor-operated data processing facilities to advise, assist, and evaluate automated systems on aspects of automated system security. Each evaluation performed by the teams identifies to management, potential vulnerabilities of the total automated operation by analyzing areas of personnel, physical, document, communication, hardware, software, procedural, and management security. The team will provide the commander or Data Processing Activity (DPA) manager with an assessment of the vulnerabilities to the system and prescribe countermeasures which must be implemented or accounted for in the risk management program.

Support to Acquisition and SAPs INSCOM provides CI support to research, development, test, and evaluation (RDTE); acquisition elements through the Acquisition Systems Protection Program (ASPP); and the safeguarding of defense systems anywhere in the acquisition process as defined in DODD 5000.1. Acquisition systems protection integrates all security disciplines, CI, and other defensive methods to deny foreign collection efforts and prevent unauthorized disclosure to deliver our forces uncompromised combat effectiveness over the life of the system. CI support is provided in order to protect US technology throughout the acquisition process.

SAPs usually involve military acquisition, operations, or intelligence activities. When applicable, CI support to SAPs extends to government and industrial security enhancement; DOD contractors and their facilities in coordination with DIS as appropriate; and the full range of RDTE activities, military operations, and intelligence activities for which DA is the proponent or executive agent. INSCOM is responsible for providing life cycle CI support to SAPs. See AR 380-381.

Support to Intelligence Disciplines CI supports the collection of HUMINT, SIGINT, and IMINT. As threats are identified and located, US Army intelligence systems are used to provide early warning, situation development, and other IEW functions. By crosswalking CI information to intelligence collection and vice versa, it eliminates possible conflict and compromise and provides a value added to the total intelligence community.

Support to Treaty Verification A security consequence of arms control is overt presence of FIS at US facilities. CI is concerned with non-treaty related activities of FIS visits to Army installations, and protecting installation activities and facilities not subject to treaty verification. CI personnel provide advice and assistance to installation commanders, and debrief Army personnel who may have come in contact with inspectors. The On-Site Inspection Agency has overall responsibility for CI support to treaty verification. INSCOM with United States Army Forces Command (FORSCOM) support is responsible for treaty verification support within CONUS. OCONUS and all CI elements provide CI support as directed by affected unified or allied command Commanders in Chief (CINCs).

Support to Domestic Civil Disturbance The primary CI function is to support unit force protection through close and continuous liaison with civilian law enforcement agencies (LEAs). Civilian LEAs are the primary information collectors and retention agencies. Since military support to civilian LEAs is a law enforcement function, EO 12333 and AR 381-10 do not apply; however, DODD 5200.27, AR 380-13, and AR 381-20 do apply. Any activity by CI personnel must comply with the following:

■   Prior to execution, CI support is coordinated with the task force senior intelligence officer and legal advisor, and is approved by the task force commander's designated law enforcement representative. Support should be confined to analytical and situation development activities.

■   CI personnel assigned to the task force work in uniform and do not use a CI badge and credentials.

To avoid questionable or illegal activities by CI personnel participating in this type of OOTW, thorough training prior to deployment is crucial. CI personnel must understand they do not conduct any activity without prior approval, and do not collect or maintain information on US persons beyond that specifically authorized for the deployment duration.

Support to Natural Disaster Operations. Without an identifiable threat to US Army security interests, use of CI personnel is not recommended since there is no viable mission for them.

Support to HUMINT Countering foreign HUMINT capabilities and efforts is a key part of accomplishing the CI mission. Again, the CI agent or CI team cannot do this task alone. CI is a total Army mission that relies on our ability to

■ Identify the hostile HUMINT collector.

■ Neutralize or exploit the collector or deny information.

■ Control our own information and indicators of operations so they are not readily accessible to foreign collectors.

■ Support C-HUMINT commanders through effective and stringent adherence to physical, information, and personnel security procedures governed by Army regulations. They apply force or assets to ensure security daily. The IEW organization provides continuous and current threat information so the command can carry out its security responsibilities.

C-HUMINT Operations C-HUMINT requires effective and aggressive offensive and defensive measures as shown in Figure 3-1. Our adversaries collect against our forces using both sophisticated and unsophisticated methods. On the battlefield we must combat all of these methods to protect

OFFENSIVE	DEFENSIVE
Targeting for fire and maneuver	Deception operations (OPSEC)
Counter espionage operations	
Counterreconnaissance	Physical security
Countersabotage	Information security
Counterterrorism	Personnel security
Penetration and exploitation operations	

Figure 3-1. C-HUMINT operations.

28

our force and to ensure the success of our operations. The CI agent recommends countermeasures developed by CI analysts that the commander can take against enemy collection activities.

To accomplish C-HUMINT, the CI agent, individually or as part of a CI team, conducts investigations, operations, and collection. A detailed description and instructions on how to conduct investigations are included in Chapter 4 and Section I to Appendix A. Other tasks include

■ Developing, maintaining, and disseminating multidiscipline threat data and intelligence files on organizations, locations, and individuals of CI interest. This includes insurgent and terrorist infrastructure and individuals who can assist in the CI mission.

■ Performing PSIs and records checks on persons in sensitive positions and those whose loyalty is questionable. Procedures for these investigations and checks are outlined in Chapter 4 and Section VII to Appendix A.

■ Educating personnel in all fields of security. A component of this is the multidiscipline threat briefing. Briefings can and should be tailored, both in scope and classification level. Briefings could then be used to familiarize supported commands with the nature of the multidiscipline threat posed against the command or activity.

■ Searching for people who pose an intelligence collection or terrorist threat to US Forces. Should CI investigations result in identifying the location of terrorists, their apprehension is done in conjunction with civil and law enforcement authorities.

■ Debriefing selected personnel (friendly and hostile) including combat patrols, aircraft pilots, or other elements which may possess information of CI interest. Individuals and types of information which may be of interest to CI personnel are identified in the paragraphs below.

■ Recognizing that the circumstances of combat and CI operations in tactical areas make the tasks of the CI agent more challenging. There may be many people who are looked upon as threats to security, perhaps solely because of their presence in the combat zone.

The CI agent must conduct preliminary screening to permit concentration on those of greatest potential interest or value. The CI agent is extremely dependent on such agencies as the MP, Civil Affairs (CA) units, Defense HUMINT Services (DHS), and interrogation prisoner of war (IPW) to identify persons of CI interest. CI personnel must keep these agencies posted on individuals of CI interest and the procedures to notify CI personnel of their detention.

Such personnel are usually apprehended as they try to enter or leave a zone of operations. If they are foreign agents, they will have cover stories closely paralleling their true environments and identities. It is, therefore, necessary that CI agents know about the AO and adjacent areas as well as the intellectual, cultural, and psychological peculiarities of the population. CI agents must also develop their linguistic and interrogation skills. Individuals who may be of interest include

- Refugees or displaced persons with knowledge of hostile or terrorist activities, intentions, locations, or sympathizers.

- Members of underground resistance organizations seeking to join friendly forces.

- Collaborators with the adversary.

- Deserters from adversary units.

- Target personalities, such as those identified on the "detain" and "of interest" lists. (See Section V to Appendix A.)

- Individuals previously holding political or civic positions of influence within the AO.

- Individuals with knowledge of the adversary force's strategic capabilities, resources, and intentions.

CI teams of technicians, agents, interrogators, and analysts not only improve the overall effectiveness of CI efforts but also expand the scope and knowledge of both. Although CI agent personnel are not interrogators, they should be knowledgeable in these areas:

■ They must be capable of expeditiously recognizing, detecting, exploiting, and reporting tactical and order of battle (OB) data.

■ All members of the CI team should know which units of the opposition their unit is facing and which units are in the AO.

■ The team should maintain current OB information for possible exploitation of EPW. Interrogators use any and all information about adversary disposition, strength, weaknesses, composition, training, equipment, activities, history, and personalities.

■ The linguistic capabilities of interrogators.

In most cases, deployment of CI agent personnel is in rear areas rather than forward with the combat units. However, CI personnel are prepared to operate whenever and wherever indications of adversary activities require a CI effort. By the very nature of their mission, CI agents provide area coverage and are in a position to provide valuable assistance to supported commands in countering adversary activities in those areas. Other missions of the CI team include

■ Developing civilian human source networks dispersed throughout the area which can provide timely and pertinent force protection information. See FM 34-5 (S).

■ Providing security advice and assistance to critical installations and activities.

■ Conducting continued briefings to MP, CA personnel, interrogators, and G2/S2 staffs on adversary intelligence activity and method of operation, as well as other threats to force protection.

■ Providing assistance to and support of the continuing program of security orientation and indoctrination of all command personnel, emphasizing the SAEDA regulation.

■ Recommending specific countermeasures to unit and installation commanders for enhancing security

31

practices, including detecting behavioral patterns detrimental to the security of the command.

- Conducting evaluations and surveys on the effectiveness of security measures.

CI personnel maintain contact and conduct continuous liaison with adjacent units and staffs. Many adjacent units have a concurrent requirement to collect intelligence on the threat, insurgent, and adversary organizations. The exchange of information is a normal function of CI personnel among adjacent units and agencies. CI personnel conduct liaison with key agencies such as MP and CA. CA personnel are great sources of initial operational, threat, and source lead information, since they are often some of the first personnel to arrive in an area. Although CA is valuable during wartime, it is even more important in OOTW. A close working relationship and continuous coordination between CA and CI elements are essential at all levels of operation and aid in the exchange of information. CA units, which have as their primary goal the reestablishment of civil order within a troubled country, can be of great assistance in the following areas:

- The identification of a subversion, espionage, or sabotage threat to US Forces.

- Thorough contact with civilian organizations, leaders, and political enemies of the government. CA units can compile personnel rosters or identify possible adversary infiltrators for screening by CI.

- Both CA and CI personnel must continuously monitor the feelings and attitudes of the civilian population. Although CI personnel orient their monitoring mission primarily on subversion, espionage, and sabotage, they must also work with CA on programs designed to counter and neutralize a hostile threat.

- CA personnel often deal with both personnel and material which can be of assistance to CI agents. Some sources which CA personnel are likely to locate may include

  - Civilians who were associated with our adversaries.
  - Leaders of fraternal, civic, religious, or patriotic organizations.

- Persons employed in recreational areas.
- Governmental documents, libraries, or archives.
- Industrial and commercial records.
- Technical equipment, blueprints, or plans.

■ Other CA personnel are in contact with leaders of civil organizations having direct contact or support activities with the military installation or population. Civil security measures requiring close coordination between CA and CI personnel include

- Population and resources control.
- Civil censorship.
- Security screening of civilian labor.
- Monitoring of suspect political groups.
- Industrial plant protection.

Coordination of efforts and the exchange of information between MP and CI agent personnel must be continuous. MP and CI agents have a mutual interest in many areas and may find themselves interfacing in a variety of circumstances. Both MP and CI elements frequently develop information which may fall into the jurisdiction of the other. The following information should be shared at the lowest practical level consistent with command policy on coordination:

■ **MP Investigations.** MP investigations are concerned with the impact of crime on the war effort. They are concerned with the ties criminals might have with local leaders, political parties, labor unions, legitimate businesses, and governmental agencies. Those criminal actions and parties may extend into the subversive and sabotage arenas. Terrorism through sabotage is a criminal act and may well be a coordinated enemy effort. Both MP and CI elements can develop offensive and defensive measures to deny the enemy information and material for hostile actions.

■ **MP and CI Elements.** MP and CI elements require coordination. MP and CI personnel jointly conduct raids, cordon and search operations, and apprehend persons of

33

CI interest. See Section VI to Appendix A. They operate joint mobile and static checkpoints for either MP control purposes or CI spot-checks for infiltrators. The intelligence exchange between these two elements is critical. The exchange may be conducted formally through staff elements, or on a liaison-type basis.

C-SIGINT Operations. C-SIGINT operations, including COMSEC monitoring (see AR 380-53) conducted during peace, war, and OOTW, are performed to enhance force protection, survivability, mobility and training; provide data

OFFENSIVE	DEFENSIVE
Targeting for fire and maneuver	Radio OPSEC countermeasures
Electronic attack	Use of secure telephone
	Signals security (SIGSEC) procedures
	Deception operations

Figure 3-2. C-SIGINT operations.

to identify friendly C-E vulnerabilities; provide countermeasures recommendations; and when implemented, determine if countermeasures are effective. C-SIGINT includes full identification of the threat and an integrated set of offensive and defensive actions designed to counter the threat as shown in Figure 3-2. C-SIGINT is addressed in greater detail in Appendix B.

C-SIGINT provides the commander with the knowledge to assess the risk and probable success of alternatives before a plan is implemented. C-SIGINT is a cyclic process requiring a strong analytical approach. The key is to be predictive. C-SIGINT is based on a thorough knowledge of

■ Foreign SIGINT.

■ Friendly force Communications-Electronics profile.

■ Foreign forces operations and plans.

■ Realistic security measures that can be taken to deny information to the enemy.

The Joint COMSEC Monitoring Activity and INSCOM MI group or brigade will provide C-SIGINT operational support at all echelons as outlined in DODD 4640.6 and NACSI 4000.

Acquiring the necessary information on foreign SIGINT collectors as well as information to support maintenance of friendly communications nodes database at all echelons is, at best, difficult. Presently, we gather adversary information from the existing SIGINT and electronic warfare (EW) collection capability of the IEW force. We also gather it through electronic preparation of the battlefield based on what we know foreign SIGINT or radio electronic combat doctrine to be.

Friendly C-E patterns and signatures information are gathered by examining our technical manuals; getting anomaly emissions information from electronics maintenance personnel; and getting procedural information from operations staffs and signal command and staff personnel.

Currently, there is no dedicated capability to accurately and completely capture friendly C-E emissions in the same way our adversaries do. Our analysis efforts attempt to make up for this shortcoming on the doctrinal portrayal of our C-E assets. At all echelons, the goals are the same:

■ Force protection.

■ Reliable use of the electromagnetic spectrum for friendly C 2.

■ Degradation or neutralization of hostile SIGINT and radio electronic combat assets.

These goals are achieved through vigorous application of a five-step C-SIGINT process:

■ Threat assessment.

■ Vulnerability assessment.

■ Recommendation of countermeasures options.

■ Countermeasures implementation.

■ Evaluation of countermeasures effectiveness.

Threat assessment must be done first if the remaining steps are to be viable. When we determine that a threat exists for a given area, the MDCI analyst assesses friendly C-E systems within that area to determine which are vulnerable to the threat. Once this has been done, the MDCI analyst develops countermeasures that will reduce or eliminate the threat, the vulnerability, or both. The commander then directs countermeasures implementation. Recommended countermeasures include

■ OPSEC (all COMSEC and electronic security measures).

■ Targeting.

■ Electronic attack (EA).

■ Electronic deception.

The last step is evaluating the measure of success for any implemented countermeasures. This may be done by

■ COMSEC monitoring.

■ Analyzing reports received.

■ Analyzing air and ground operational situation reports.

■ Analyzing of data collected during or after countermeasures implementation.

OFFENSIVE	DEFENSIVE
Action of ADA	Deception operations
	OPSEC countermeasures
Targeting for fire and maneuver	Aerial platform tracking

Figure 3-3. C-IMINT operations.

C-IMINT Operations C-IMINT is a total force mission that includes full identification of the adversary and an integrated set of actions to counter the threat. These actions range from offensive action to the use of OPSEC techniques and deception designed to deny adversaries' information as shown in Figure 3-3.

Threat. As with all CI operations, detailed analysis of the adversary is necessary. To help protect our force from exploitation, our analysts must fully understand the adversary and its capabilities.

Any adversary may possess or acquire systems or products with a comprehensive and sophisticated IMINT capability. We must have in place a carefully developed counterprogram to negate any tactical and strategic threat. An adversary or belligerent acquires IMINT information through a variety of ways from handheld cameras to sophisticated satellite reconnaissance systems. Adversary IMINT systems may include

■ Aerial cameras.

■ Infrared sensors.

■ Imaging radars.

■ Electro-optical sensors (TV).

■ Multispectral and digital imagery products.

Adversaries know that to maximize the effect of their massed fire-power and mobility, their IMINT capabilities must be accurate and timely. They know their IMINT operations will be met by our countermeasures and deception operations. For this reason, they may use diverse multi-sensor collection means to obtain information. Sometimes the various collection means and sensors overlap and are redundant.

Adversary commanders and staffs of all combat arms and services organize reconnaissance operations. Adversary doctrine indicates that reconnaissance is effective only if it is actively and continuously conducted under all conditions and circumstances. Continuity of action, timeliness, and accuracy of information are constantly stressed.

Tactical air reconnaissance is a good source of IMINT. Adversaries use air reconnaissance at all levels with organic or supporting manned and unmanned aviation assets.

Adversary IMINT collection efforts directed against US and allied forces vary according to weather, terrain, and the depth and density of friendly forces and their collection capabilities. Reconnaissance aircraft, in general, also carry weapons and are capable of attacking ground targets of opportunity.

C-IMINT Collection. We must view a potential adversary's use of IMINT to develop intelligence and targeting information as potentially damaging to our interests. We get information on adversary IMINT operations from many different sources:

■ Enemy prisoner of war (EPW) reports.

■ US Air Force reports.

■ Tactical and strategic reconnaissance.

■ HUMINT operations.

■ SATRAN. See DIAM DJS-1400-7-85 (S).

■ Air defense artillery (ADA) reports.

■ SPOT reports in size, activity, location, unit, time, and equipment (SALUTE) format.

■ CI threat assessments, estimates, and summaries from higher, lower, and adjacent units.

■ OPSEC surveys, estimates, and assessments.

Techniques

CI techniques are means used to accomplish the mission efficiently and effectively. Selection of techniques occurs at the lowest level possible by the on-scene CI element to meet the needs of the supported military commander within the constraints of the operation and applicable regulations. Techniques include vulnerability assessments, hostile intelligence simulation (Red Team), and covering agent support.

Vulnerability Assessments:

Vulnerability assessments are studies conducted by CI personnel to provide a supported command or agency a picture of its susceptibility to foreign intelligence collection. These assessments may be conducted on a command, agency, installation, subordinate element, HQ, operation, or program and are tailored to the needs of each requestor. The objective is to provide a supported command or agency a realistic tool with which to evaluate internal force protection or security programs, and to provide a decisionmaking aid for the enhancement of these programs. Vulnerability assessments include

■ Evaluating FIS multidiscipline intelligence collection capabilities, collection and other activities, and PIR.

■ Identifying friendly activity patterns (physical and electronic), friendly physical and electronic signatures, and resulting profiles.

■ Monitoring or collecting C-E transmissions to aid in vulnerability assessments, and providing a more realistic and stable basis from which to recommend countermeasures.

■ Identifying vulnerabilities based upon analysis of collected information and recommendations of countermeasures.

■ Analyzing the effectiveness of implemented countermeasures.

Hostile Intelligence Simulation (Red Team):

Upon request by a commander or program manager, CI personnel may plan and execute a simulation of a foreign intelligence penetration of a specified target, such as an installation, operation, or program. Such simulations are informally known as Red Team operations. There is no single structure or composition for a Red Team. It is an array of MDCI resources which are selectively employed during the operation. Red Team operations include the full range of MDCI activities to include those activities formerly performed as vulnerability assessments that may be applied to replicate the FIS threat.

Red Team operations provide a supported command or agency a tool to evaluate internal force protection or security programs, and a decisionmaking aid for the enhancement of these programs. Red Team operations assist the commander or program manager and his security staff by identifying vulnerabilities based upon analysis of the collected information and recommending countermeasures to reduce or negate those vulnerabilities.

Red Team operations should be carried out as realistically as possible in accordance with AR 381-10 and AR 381-20. Red Team operations should be conducted by the most experienced CI personnel available after thorough coordination with the unit commander and security manager. These operations require extensive preparation. A full multidiscipline Red Team operation would require support from EAC CI elements.

Commanders must ensure compliance with laws, policy, and regulations when employing COMSEC monitoring, electronic surveillance, or other technical CI collection activities as part of Red Team simulation operations.

Because of the complexity and high resource requirements, Red Team operations generally should be limited to extremely sensitive activities, such as SAPs, although Red Team operations may be useful in conjunction with major tactical exercises and deployments. For more information on Red Team operations, see AR 381-20.

Red Team proposals will be documented in an OPLAN and approved by the activity head or commander who requested the service. Red Team findings will be used to inform and educate commanders and their security staffs on the effectiveness of their security policies and practices. CI personnel also assist the command in enacting countermeasures to any vulnerabilities detected by Red Team operation.

Covering Agent Support:

CI covering agent support is the technique of assigning a primary supporting special agent to a command or agency. This agent will conduct all routine liaison and advice and assistance with the supported element. It ensures detailed familiarity with the supported element's operations, personnel, security, and vulnerabilities, and in turn provides the element with a point of contact for reporting matters of actual or potential CI interest.

Chapter 4

Counterintelligence Collection Activities

General

CI collection activities gather the information the commander needs to make decisions in support of the overall mission. CI activities help the commander shape the battlefield. The commander focuses the CI effort by carefully assigning missions and clearly defining the desired results. By orienting the unit's CI capabilities, the commander decides who or what are CI targets for collection activities. This chapter describes sources of CI information, control of sources, CI liaison, and touches on debriefings and CFSO.

CI agents conduct CI collection operations in support of the overall mission. CI agents are augmented by interrogators when they are available. These operations rely on the use of casual as well as recruited sources of information to satisfy specific requirements of a command or activity supported by CI. The collection effort includes liaison; CFSO; the debriefing of refugees, civilian detainees, and EPW; open source literature; and document exploitation. These operations use the techniques identified in FM 34-5 (S). AR 381-172 (S) covers the policy concerning CFSO. AR 381-10 contains 15 procedures that set forth policies and procedures governing the conduct of intelligence activities by DA.

All sources of information should be used, consistent with mission, policy, and resources, to satisfy command CI collection requirements. Several sources of information are discussed below:

- A casual source is one who, by social or professional position, has access to information of CI interest, usually on a continuing basis. Casual sources usually can be relied on to provide information which is routinely available to them. They are under no obligation to provide information. Casual sources include private citizens, such as retired officials or other prominent residents of

41

an area. Members of private organizations also may furnish information of value.

- Official sources are liaison contacts. CI personnel conduct liaison with foreign and domestic CI, intelligence, security, and law enforcement agencies to exchange information and obtain assistance. CI personnel are interested in investigative, operational, and threat information. See CI Liaison below.

- Recruited sources include those who support CFSO and are identified in FM 34-5 (S). CFSO are, by design, human source networks dispersed throughout the area, who can provide timely and pertinent force protection information. See FM 34-5 (S) and CI Force Protection Source Operations below.

- Refugees, civilian detainees, and EPWs are other sources of CI information. Interrogators normally conduct these collection operations, often with technical assistance from a CI agent. The key to identifying the source of valuable CI force protection information is in analyzing the information being sought and predicting who, by virtue of their regular duties, would have regular, frequent, and unquestioned access to such information.

- Open source publications of all sorts and radio and television broadcasts are valuable sources of information of CI interest and operational information. When information is presented in a foreign language, linguist support is required for timely translation. Depending on the resources, this support can be provided by interrogation personnel, allied personnel, indigenous employees, or Reserve Component (RC) translators (97L).

- Documents not openly available, such as adversary plans and reports, are exploited in much the same way as open source publications.

Control of Source Information

All collection operations require keeping records on sources of information. This holds true for liaison contacts as well as casual or recruited sources. These types of operations require security and maintenance of source information in intelligence operations channels. This helps to preclude any compromise of sources or friendly methods of operation. This type of information, including biographic, motivational, and communications procedures, are best maintained in CI C 2 channels. Control of source information will not preclude passage of this type of information from one echelon to another for necessary approvals.

In handling source information, strictly adhere to the "need-to-know" policy. The number of persons knowing about source information must be kept to a minimum. For more information on the control of source information and CI collection activities, see FM 34-5 (S).

CI Liaison

CI agents conduct CI liaison to obtain information, gain assistance, and coordinate or procure material. The nature of CI activities and the many legal restrictions imposed, including SOFAs or other agreements, make the collection of intelligence information largely dependent on effective liaison. CI agents use liaison to obtain information and assistance and to exchange views necessary to understand our liaison counterparts. During transition from increased tension to open hostilities, the liaison emphasis shifts to support the combat commander. CI agents must establish liaison with appropriate agencies before the outbreak of hostilities. Information and cooperation gained during this period can have a major impact on the effectiveness of both intelligence and combat operations. Liaison with foreign organizations and individuals normally requires foreign language proficiency.

Liaison with appropriate US, host country, and allied military and civilian agencies is fundamental to the success of CI operations and intelligence support to commanders. In many cases, full-time liaison officers (LNOs) or sections are necessary to maintain regular contact with appropriate organizations and individuals. In addition to national agencies, numerous local agencies and organizations also provide assistance and information.

A basic tenet of liaison is quid pro quo (something for something) exchange. While the LNO sometimes encounters individuals who cooperate due to a sense of duty or for unknown reasons of their own, an exchange of information, services, material, or other assistance normally is part of the interaction. The nature of this exchange varies widely, depending on location, culture, and personalities involved.

The spectrum of liaison tasks ranges from establishing rapport with local record custodians to coordinating sensitive combined operations at the national level of allied nations. Commanders with CI assets involved in liaison should provide the following guidance:

■ **Liaison objectives.** Liaison objectives are types of information to be collected, methods of operations unique to the area, and command objectives to be accomplished.

■ **Limitations on liaison activities.** These limitations include

   • Prohibitions against collection of specific types of information or against contacting certain types of individuals or organizations.

   • Memorandums of Understanding with other echelons delineating liaison responsibilities.

   • Delineation of areas of responsibility of subordinate elements.

   • Director of Central Intelligence Directives (DCID).

■ **Administrative considerations.** Some administrative considerations include

   • Type, method, and channels of reporting information obtained from liaison activities.

   • Project and intelligence contingency fund cite numbers to be used.

   • Funding and incentive acquisition procedures.

   • Limitations on the use of intelligence contingency fund or incentives.

   • Budget restraints.

   • Source coding procedures, if used.

   • Report numbering system.

   • Procedures for requesting sanitized trading material information.

- **Authority.** Authority under which the specific liaison program is conducted and guidelines for joint and combined operations are set.

- **Other.** Other SOPs cover related aspects, such as funding, intelligence information reporting procedures, source administration, and areas of responsibility and jurisdiction.

In CONUS, CI liaison provides assistance in operations and investigations, precludes duplication of effort, and frequently provides access to information not available through other CI channels. Agents should maintain a point of contact roster or list of agencies regularly contacted. Agencies normally contacted on a local basis include

- Military G2, S2, and personnel sections of units in the area.

- G5 and S5 representatives.

- MP and provost marshal.

- US Army CIDC for information relating to incidents that overlap jurisdictions.

- Civilian agencies such as state, county, or local police departments; state crime commissions; state attorney general offices; and local courts.

- Local offices of federal agencies such as the FBI, Immigration and Naturalization Service (INS), Border Patrol, Drug Enforcement Agency, and similar security agencies.

- Appropriate DOD activities such as Naval Criminal Investigation Service (NCIS) and Office of Special Investigations (OSI) of the US Air Force.

The Office of DCSINT is responsible for liaison with the national headquarters of the intelligence community and other agencies for policy matters and commitments. CG, INSCOM, is the single point of contact for liaison with the FBI and other federal agencies for coordinating operational and investigative matters.

Overseas CI liaison provides support to a number of diverse US Government agencies. This support ranges from conducting tactical operations to fulfilling national level requirements generated by non-DOD federal agencies. Individuals contacted may include private individuals who can provide assistance, information, and introductions to the heads of national level host country intelligence and security agencies. Overseas liaison includes the overt collection of intelligence information.

Compared to the US, many countries exercise a greater degree of internal security and maintain greater control over their civilian population. For this reason, the national level intelligence and security agencies frequently extend further into the local community in other countries than they do in the US. Security agencies may be distinctly separate from other intelligence organizations, and police may have intelligence and CI missions in addition to law enforcement duties. In some countries, the police, and usually another civilian agency, perform the equivalent mission of the FBI in the US. This other civilian agency frequently has a foreign intelligence mission in addition to domestic duties. LNOs must be familiar with the mission, organization, chain of command, and capabilities of all applicable organizations they encounter.

Operational benefits derived from CI liaison include

- Establishing working relationships with various commands, agencies, or governments.

- Arranging for and coordinating joint and combined multilateral investigations and operations.

- Exchanging operational information and intelligence within policy guidelines.

- Facilitating access to records and personnel of other agencies not otherwise available. This includes criminal and subversive files controlled by agencies other than MI. Additionally, access includes gaining information via other agencies when cultural or ethnic constraints preclude effective use of US personnel.

- Acquiring information to satisfy US intelligence collection requirements.

Language proficiency is a highly desirable capability of a CI agent conducting liaison. It is easier to deal with a liaison source if the LNO can speak directly to the source rather than speak through an interpreter. Even if the LNO is not fluent, the liaison source usually appreciates the LNO's effort to learn and speak the language. This often enhances rapport.

Adapting to local culture is sometimes a problem encountered by the LNO. Each culture has its own peculiar customs and courtesies. While they may seem insignificant to US personnel, these customs and courtesies are very important to local nationals.

Understanding a country's culture and adhering to its etiquette are very important. What is socially acceptable behavior in the US could very well be offensive in other cultures. Knowing the local culture helps the LNO understand the behavior and mentality of a liaison source. It also helps in gaining rapport and avoiding embarrassment for both the liaison source and the LNO. In many cultures, embarrassing a guest causes "loss of face." This inevitably undermines rapport and may cause irreparable harm to the liaison effort.

The LNO also must understand the capabilities of agencies other than our own. Knowledge of the liaison source's capabilities in terms of mission, human resources, equipment, and training is essential before requesting information or services. Information exchanged during the conduct of liaison is frequently sanitized. Information concerning sources, job specialty, and other sensitive material relating to the originator's operations may be deleted. This practice is common to every intelligence organization worldwide and should be taken into account when analyzing information provided by another agency.

The LNO may have to deal with individuals who have had no previous contact with US agencies and who are unsure of how to deal with a US intelligence agent. The LNO must remember that to the liaison source, they represent the people, culture, and US Government. The liaison source assumes the behavior of the LNO to be typical of all Americans. Once the American identity becomes tarnished, it is difficult for the LNO, as well as any other American, to regain rapport.

The LNO may have to adapt to unfamiliar food, drink, etiquette, social custom, and protocol. While some societies make adjustments for an "ignorant foreigner," many expect an official visitor to be aware of local customs. The LNOs must make an effort to avoid cultural shock when confronted by situations completely alien to his background. The LNO also must be able to adjust to a wide variety of personalities.

Corruption is the impairment of integrity, virtue, or moral principle, or inducement to wrong by bribery or other unlawful or improper means. In some countries, government corruption is a way of life. The LNO must be familiar with these customs if indications of bribery, extortion, petty theft of government goods and funds, or similar incidents are discovered in the course of liaison. When corruption is discovered, request command guidance before continuing liaison with the particular individual or organization. Regardless of the circumstances, exercise caution and professionalism when encountering corruption.

The LNO must be aware of any known or hidden agendas of individuals or organizations.

Jealousy between agencies is often a problem for the LNO. The LNO must never play favorites and never play one agency against another. Occasionally, due to the close professional relationship developed during liaison, a source may wish to present a personal gift. If possible, the LNO should diplomatically refuse the gift. If that is not possible, because of rapport, accept the gift. Any gifts received must be reported in accordance with AR 1-100. The gift can be kept only if you submit and get approved a request to do so. The same restrictions also apply to the LNO's family.

Records and reports are essential to maintain continuity of liaison operations and must contain information on agencies contacted. It is preferable to have a file on each organization or individual contacted to provide a quick reference concerning location, organization, mission, and similar liaison-related information. Limit information to name, position, organization, and contact procedures when liaison is a US person. For liaison contacts with foreign persons, formal source administrative, operational, and information reporting procedures are used. Guidance for these procedures is in FM 34-5 (S).

Debriefing

Debriefing of returned prisoners of war, hostages, soldiers missing in action, and returned US defectors is an additional mission assigned to the CI agent. The purpose of these debriefings is to

■ Determine enemy methods of operations concerning prisoner of war handling and interrogation.

■ Learn of enemy weaknesses.

■ Gain information concerning other prisoners and soldiers missing or killed in action.

■ Conduct a damage assessment.

■ Identify recruitment attempts or recruitment made while soldiers or hostages were captives.

■ Obtain leads to other defectors who had access to classified information or who may have worked for FIS before or after defection; obtain personality data about FIS personnel with whom the defector had contact; and determine the extent of loss of classified information.

CI Force Protection Source Operations

CFSO evolved out of low-level source operations (LLSO), defensive source operations (DSO), and tactical agent operations (TAO). LLSO are still accomplished by non-CI teams charged with these types of missions. See FM 34-5 (S). The change in terminology was a result of the push for similar terminology amongst separate service activities responsible for source operations and to tie these operations directly to the force protection support needs of the combat commander.

CFSO support force protection of deployed US Forces and are governed by AR 381-172 (S). CFSO are conducted when directed by the theater CINC or Army component commander or their senior intelligence officers, and are conducted and supervised by CI officers and agents with appropriate linguist support from HUMINT collectors-interrogators.

CFSO fill the intelligence gap between the needs of the combat commander and national level requirements. These operations are designed to be both aggressive and flexible in nature to quickly respond to the needs of the supported command. CFSO are focused to collect force protection information on local terrorists, saboteurs, subversive activities, and other hostile activities affecting the security of deployed US Forces.

All considerations listed previously in CI Liaison involving liaison contacts, specifically language proficiency, local customs, and capabilities are germane to CFSO. For more information on CFSO, see AR 381-172 (S) and FM 34-5 (S).

Chapter 5

Counterintelligence Analysis and Production

General

Analysis and production is the heart of intelligence. No matter what quality and quantity of information is gathered, it does absolutely no good if the information is not turned into intelligence and disseminated to the commander in time for him to use it in the decisionmaking process. The same is doubly true of CI. CI agents, interrogators, and MDCI analysts work in teams to gather information, process it into intelligence, put it into products usable at all levels, and disseminate it in time to keep our commander's decision time inside the decision time required by an adversary.

CI analysis and production is focused on three well-defined FIS activities: HUMINT, SIGINT, and IMINT. The process of countering each of these disciplines involves a threat assessment, vulnerability assessment, development of counter-measures options, countermeasures implementation, and countermeasures evaluation. These are referred to as the five-step CI process. (See Section II through Section V of Appendix B.) But they are more than that.

■ While each step is a product, it is also a process. Each step can stand alone, yet each depends upon the other for validity. Once begun, the five-step CI process becomes cyclic. The cyclic process does not end, for within each step is the requirement for continuous updating of the CI database. This is necessitated by any new information reflecting change in either the FIS posture, the friendly posture, or both.

■ Because FIS activities involve collection, analysis, and production and are themselves multidisciplined, efforts to counter FIS activities will likewise be multidisciplined and will require collection, analysis, and production in order to be successful. The analyst will be able to produce a truly multidisciplined product only if collection is productive.

- Collection is a single discipline function and the atten-
  dant initial analysis is likewise a single discipline. The
  fusion and refined analysis of individual disciplines
  occurs at various echelons of command, specifically the
  ACE at theater, corps, and division and at the Army CI
  Center, 902d MI Group, Fort Meade, MD.

CI analysis is by no means exclusive to Army agencies, but
is a crucial activity of DOD. CI analysis is performed at the
Defense Intelligence Agency (DIA), as well as other federal
agencies such as the Central Intelligence Agency (CIA), FBI,
and the National CI Center. CI analysis must be performed by
highly trained, experienced, and skilled analysts using the lat-
est technology and modern methods of planning and directing, processing, producing, and disseminating.

C-HUMINT:

HUMINT analysis focuses not only upon the FIS entity or
entities operating in the area but also upon the intelligence
product most likely being developed through their collection
activities. The analytical effort should attempt to identify the
FIS HUMINT cycle (collection, analysis, production, targeting) and FIS personalities. To produce a complete product,
the MDCI analyst may need access to considerable data and
require significant resources. The MDCI analyst will require
collection in the areas of subversion, espionage, sabotage,
terrorism, and other HUMINT supported activities. Collection
of friendly data is also required to substantiate analytical
findings and recommendations. Consistent with time, mission,
and availability of resources, efforts must be made to provide
an analytical product that identifies FIS efforts.

C-SIGINT:

SIGINT like C-HUMINT focuses upon the FIS entities
which can collect on friendly forces. It also focuses on the
intelligence which is most likely being collected. Also like
C-HUMINT and C-IMINT, any C-SIGINT analysis effort should
be fully automated (data storage, sorting, and filing). The
MDCI analyst requires SIGINT data collection to support
vulnerability assessment and countermeasures evaluation.
Validation of vulnerabilities (data capturable by FIS SIGINT)
and the effectiveness of implemented countermeasures (a

before and after comparison of electromagnetic signatures and data) will be nearly impossible without active and timely collection as a prerequisite to analysis. The MDCI analyst requires a comprehensive, relational database consisting of FIS SIGINT systems, installations, methodology, and associated SIGINT cycle data. In addition, all friendly C-E systems and user unit identification must be readily available, as well as a library of countermeasures and a history of those previously implemented countermeasures and results. Ideally, the MDCI analyst should, at any given time, be able to forecast FIS SIGINT activity. However, such predictions must rely upon other CI, interrogator, SIGINT, and IMINT collection as well as access to adjacent friendly unit CI files. Information on FIS SIGINT must be readily accessible from intelligence elements higher as well as lower in echelon than the supported command.

C-IMINT:

IMINT requires the analyst to have an in-depth knowledge of the supported commander's plans, intentions, and proposed AO as far in advance of commitment as possible. The analyst must have access to all available data and intelligence on FIS IMINT methodology, systems, and processing as well as indepth information on commercial satellite systems and their availability to the foreign consumer. The analyst attempts to define the specific imagery platform deployed against US Forces and the cycle involved (time based) from time of imaging through analysis to targeting. Knowledge of FIS intelligence cycle to targeting is critical in developing countermeasures to defeat, destroy, or deceive FIS IMINT. For ground-based HUMINT oriented IMINT (video cassette recorders [VCRs], cameras, host nation curiosity, news media organizations) the CI team will be required to collect the data for the analyst. This type of information cannot be reasonably considered to exist in any current database.

Traditional FIS IMINT data is readily available and should not require any CI collection effort. However, collection to support CI (overflights of friendly forces by friendly forces) during identified, critical, and IMINT vulnerable times will validate other CI findings and justify countermeasures. This "collection" will be of immense value to the analyst and the supported commander in determining what, if anything, FIS imagery has captured. It must be done within the established or accepted FIS activity cycle.

The CI analyst uses the tools and skills identified in this chapter and in FM 34-3. The intelligence analyst focuses on "how we see the opposition"; the MDCI analyst focuses on this and "how the opposition sees us." The MDCI analyst must also focus on how to counter the opposition's collection efforts. Where the intelligence analyst is a subject matter expert on the opposition, the MDCI analyst, in addition to having an indepth understanding and expertise on foreign intelligence collection capabilities, must have a good working knowledge of our own force. The CI analysis assets of the ACE must be fully integrated into the ASAS as well as the single-source C-HUMINT processor. They require access to all-source data that is applicable to CI analytical products.

The principles and techniques identified in FM 34-3 apply equally in CI analysis. This chapter focuses specifically on the application of analysis on CI matters.

CI Analysis

The CI and C-HUMINT multidiscipline assets of the ACE are under the staff supervision of the G2 at theater, corps, and division levels. Theater ACE staffing is provided from the operations battalion of the theater MI brigade. Corps ACE staffing is provided from the corps MI brigade headquarters and operations battalion. Division ACE staffing is provided by personnel assigned to the headquarters company of the divisional MI battalion. In addition to CI personnel, an all-source mix of single discipline analysts is sometimes required for interpretation to produce the CI analytical products required by the commander at each echelon. CI products are also critical to the function of the G3 OPSEC and deception cells as well.

The CI mission is a diverse and all-encompassing CI analytical effort. MDCI analysts perform the following functions:

- Analyze the multidiscipline intelligence collection threat targeted against friendly forces.

- Assess opposition intelligence collection threat vulnerabilities and susceptibilities to friendly deception efforts.

- Support friendly vulnerability assessment.

■ Develop, evaluate, and recommend countermeasures to the commander. These countermeasures reduce, eliminate, or take advantage of friendly force vulnerabilities.

■ Support rear operations by identifying collection threats to rear area units and installations, to include low-level agents responsible for sabotage and subversion.

■ Nominate targets for exploitation, neutralization, or destruction.

■ Develop and maintain a comprehensive and current CI database.

■ Identify information gaps in the form of intelligence requirements and provide requirements to the collection management element. This element will task collection missions to the appropriate supporting MI element or request information from higher echelons.

Specific responsibilities pertaining to analysis of FIS use of HUMINT, SIGINT, and IMINT follow:

■ C-HUMINT analysis includes—
   • Analyzing and assessing the espionage, terrorism, subversion, treason, sedition, and sabotage threats.

   • Analyzing enemy HUMINT collection capabilities and activities, and further analyzing how those collection capabilities can affect the friendly command.

   • Analyzing Level I threats such as enemy controlled agents or partisan collection, and Level II threats such as diversionary and sabotage operations conducted by unconventional forces.

   • Recommending countermeasures and deception.

   • Nominating targets for exploitation, neutralization, or elimination.

■ C-SIGINT analysis includes—
   • Analyzing and assessing foreign SIGINT collection capabilities and activities.

   • Comparing opposition collection systems capabilities against friendly targets.

55

- Identifying, analyzing, and assessing friendly electronic patterns and signatures.

- Analyzing friendly vulnerabilities against foreign SIGINT collection efforts.

- Recommending countermeasures and deception.

- Nominating enemy SIGINT targets for exploitation, neutralization, or destruction.

■ C-IMINT analysis includes—
  - Analyzing and assessing adversary imagery collection capabilities and activities, to include ground, air, and space systems. Threat systems include anything from hand-held cameras to satellite platforms, or fixed or rotary-wing aircraft and unmanned aerial vehicles (UAVs). The assessment should include adversary access to commercial satellite imagery and the ability to properly analyze the imagery.

  - Measuring enemy collection systems against friendly targets.

  - Identifying, analyzing, and assessing friendly patterns, signatures, and vulnerabilities for subsequent development and recommendation of countermeasures and deception.

  - Nominating opposition IMINT systems for exploitation, neutralization, or destruction.

Other intelligence support to CI analysis cannot be conducted without the support of all three intelligence disciplines—HUMINT, SIGINT, and IMINT. These disciplines collect critical information on adversary collection, analysis, and dissemination systems. Analysts extract information from the all-source database within the ACE to determine adversary collection capabilities and operations. These systems, coincidentally, collect a great deal of intelligence on friendly forces. This intelligence is vital in evaluating friendly profiles and thereby determining their vulnerabilities. If the situation warrants, we can task friendly collection systems to specifically collect information on friendly forces for the MDCI analysts through the collection management team.

The CI mission mandates a wide range of functions and tasks that are accomplished in peace and at all intensities of conflict. CI operational activities perform such functions as investigations, operations, and collection. Their products are of great value to the MDCI analyst. MDCI analysts work with CI teams and the collection management team in the ACE, and maintain rapport with operational CI and interrogation personnel in the AO in order to obtain information from all echelons.

CI Analysis Target Nominations

The G2 nominates targets that enable the commander to exploit, neutralize, or destroy the enemy. The ability to recommend target nominations to the G2 is one of the most important contributions of the CI interrogator or CI teams.

The CI team develops target nominations by using MDCI products developed in the analytical process. Target nominations are coordinated through the single-source analysis section and all-source intelligence section for inclusion on the G2's list of high-value targets (HVTs). The commander, G3, G2, and fire support coordinator comprise an informal targeting team that develops the high-payoff target (HPT) list from the list of HVTs through the wargaming process. HPTs are those that must be acquired and attacked in order to ensure success of friendly operations. Approved targets are listed in CI threat assessments and briefings, in accordance with the unit SOP. The G2 or G3 disseminates decisions concerning actions to be taken with regard to targets. For detailed information on the operations of the ACE, see FM 34-25-3. For more information on the targeting process, see FM 6-20-10.

The commander directs that HPTs be countered in three ways— exploitation, neutralization, or destruction. Targets for exploitation are monitored for their value to friendly operations. Targets for neutralization may be isolated, damaged, or otherwise rendered ineffective so they cannot interfere with the success of friendly operations. Targets for destruction are killed.

Exploited targets can assist commanders in securing their forces and operations; and identifying windows of operational risk, areas of operational risk, and windows of operational advantage.

Exploited targets can be a combat multiplier. Exploitation should be used when the opposition element or resource can be manipulated, controlled, or in some other manner used to the advantage of the friendly force. This usually occurs when

57

the identity, capability, location, and intentions of the target are known. Key considerations in nominating targets for exploitation include

■ Friendly forces' ability to deceive, control, or manipulate the target.

■ Neutralization or destruction which is not possible or practical.

■ Exploitation which will benefit friendly forces.

■ Benefits to the friendly force which outweigh neutralization or destruction.

Targets should be neutralized when the opposition elements or resources are known and located by the friendly force, and can be rendered ineffective. Actions taken to neutralize targets can be offensive or defensive measures which prevent the opposition from achieving its objective. Usually, destruction or elimination of these targets is neither possible nor practical.

Key considerations in nominating targets for neutralization include—

■ Friendly forces' inability to destroy or eliminate the target.

■ Knowledge or ability to know the target's location, identity, capability, and intentions.

■ Friendly operational activities and resources targeted by the opposition.

■ Ability of friendly forces to neutralize the target.

Targets which may be considered for neutralization are—

■ Targets which can be effectively jammed.

■ Targets which can be isolated from their objectives through the use of physical obstacles, including barriers, friendly maneuver, and entrapment.

■ Known opposition collectors against which friendly force countermeasures can be implemented.

Countermeasures developed to neutralize a target are specific measures in addition to OPSEC measures. This may include moving a tactical operations center (TOC) during a known window of advantage; working with the G4 to redesignate main supply routes based on a known threat; and recommending barrier locations to engineers. Remember, nominating targets for destruction or elimination is almost always preferable to nominating targets for neutralization or recommending actions to neutralize targets—provided destruction of the target is practical.

Destruction or elimination of targets. These targets are battalion size or smaller, which the friendly force can destroy or render combat ineffective or render intelligence collection ineffective. Usually, the identity, capability, intentions, and locations are known. Targets which may be recommended for destruction include

■ Bases of airborne reconnaissance units.

■ Hostile intelligence services operatives, saboteurs, and terrorists.

■ Base camps for opposition unconventional warfare forces either in friendly or opposition territory.

■ Special purpose forces.

■ The entire spectrum of enemy intelligence collection, analysis, and dissemination systems, including critical enemy command posts (CPs).

MP, special reaction forces, attack helicopters, field artillery, tactical air, or infantry can destroy targets. When nominating for destruction targets which are located behind friendly lines, the analyst must consider the risk to friendly forces in making the recommendation. Exact CI analysis, fully coordinated with the G3, is essential. If all the necessary information for destruction of the target with minimal risk to friendly forces is not available, it may be better to recommend neutralization of the target.

Consider:

■ The ability of friendly forces to destroy the target without undue risk.

59

- The ability to isolate, locate, and identify the target.

- The availability of friendly forces to accomplish destruction.

- The destruction is beneficial to friendly forces.

- The overall gains outweigh potential risks.

The end purpose of all analysis is to enable the friendly commander to engage and destroy the opposition. Where the opposition cannot be destroyed outright, the friendly commander must be able to exploit or neutralize them. This must be accomplished with minimal loss of friendly forces. CI analysis contributes to the accomplishment of each of these missions.

CI Analysis Products

CI analysis products convey the essence of the CI analysis to the commander and staff and higher, lower, and adjacent units. MDCI analysts prepare C-HUMINT, C-SIGINT, and C-IMINT products that become the analytical tools used to produce collective CI products. CI products also provide OPSEC or deception planners critical information required for their operations. Among these products are rear operations IPB; MDCI summaries (MDCISUMs); CI threat assessments; CI situation overlays; and CI estimates.

Rear Operations IPB:

In every operation, someone has to watch the back door. That someone is the MDCI analyst. Working in the rear CP or the combat service support CP, the MDCI analyst works through the steps of the IPB process taking a slightly different approach than his counterparts in the main CP. Specific responsibilities follow:

- MDCI analysts use maps at a scale of 1:50,000 or larger (1:25,000 scale or town plans at 1:12,500 scale are even better). This scale permits them to obtain the resolution needed to precisely locate and evaluate terrain suitable for Level I or II threats.

- MDCI analysts identify the most probable area for a small threat insertion of perhaps 6 to 10 personnel. They

60

also identify a Level III threat. Insertion of a Level III threat in the rear area would most likely take place as a cross-forward line of own troops (FLOT) operation. Close coordination with ACE analysts ensures the inclusion of IPB products to predict this threat.

- Divisional analysts are concerned with the division rear area up to the brigade rear area.

- Corps analysts would concentrate on the corps rear area down through the division rear area.

- EAC ACE is concerned with the communications zone down through the corps rear area.

- FORSCOM J2 is responsible for CONUS rear operations IPB.

During peacetime, the MDCI analyst builds an extensive database for each potential area in which threat intelligence collectors or battalion size or smaller units might operate. He analyzes this intelligence base in detail to determine the impact of enemy, weather, and terrain on operations and presents it in graphic form. The analysis has the added ingredient of assisting in the assessment of friendly COAs from the enemy's perspective. Graphics assist the commander in identifying targets as they enter the battle area. Because rear operations IPB targets consist of small units or threat intelligence collection resources, these targets are not as prominent as those viewed in the all-source products.

However, the process still generates HVTs and HPTs. Additionally, rear operations IPB assists in determining friendly HVTs and HPTs from the enemy's perspective. These are the friendly critical nodes or clusters susceptible to enemy collection or hostile action that are deemed critical to successful operations. Rear operations IPB and IPB threat evaluation use the same analytical technique—templating. Rear operations IPB templates are similar to IPB templates in the main battlefield area. They provide a comparative intelligence database for integrating threat intelligence collection activities and small unit operations with the weather and terrain for a specific area. This enables the MDCI analyst to graphically portray enemy intelligence collection and small unit

61

capabilities; depict probable COAs both before and during the battle; and confirm or refute predictions.

Both rear operations IPB templates and IPB templates are dynamic and require continual review. Not only do they portray enemy intelligence elements and small unit character-istics but they also seek to graphically portray named areas of interest (NAIs). Like the IPB process, rear operations IPB develops and employs doctrinal, situational, and event tem-plates, and matrices that focus on intelligence collection and identifying which COA an adversary will execute. These COA models are products the staff will use to portray the threat in the decisionmaking and targeting process.

MDCI analysts develop and maintain templates through-out the IPB process and provide the basis for collection and further CI analysis. The analyst's ultimate goal is the nomina-tion of targets for exploitation, neutralization, suppression, harassment, and destruction. For more information on IPB, see FM 34-130.

MDCI Summary:

The MDCISUM is a graphic portrayal of the current situ-ation from a CI point of view. The MDCI analyst uses the MDCISUM to show known adversary collection units, as well as Levels I and II threats within the friendly area. The MDCISUM is a periodic report usually covering a 12-hour period. It shows friendly targets identified as adversary objec-tives during the specified timeframe as shown in Figure 5-1. The MDCI analyst includes a clear, concise legend on each MDCISUM showing the time period, map reference, and symbols identifying friendly and adversary information. As the MDCI analyst identifies a friendly critical node, element, or resource as an adversary combat or intelligence collection target, he puts a box around it and labels it with a "T" number. The legend explains whether the "T" is—

■ A combat intelligence target.

■ A source and time confirmation.

■ An adversary resource or element that will attack or col-lect against the target in the future.

■ The expected timeframe for the adversary to exploit the target.

62

The MDCISUM incorporates rear operations IPB products and individual and specific products to the extent they are relevant to the MDCISUM reporting period. The MDCISUM might portray the following information:

■   Satellite or tactical reconnaissance patterns over the friendly area.

■   Sweeps by enemy side looking airborne radar (SLAR) or EA air platforms to the full extent of their maximum ranges.

■   Suspected landing zones or drop zones which will be used by an enemy element in the rear area.

■   Area or unit which has received unusual enemy jamming.

■   Movement of an enemy mobile SIGINT site forward along with a graphic description of the direction and depth of its targeting.

■   Location of an operational enemy agent or sabotage net.

■   Last known location of threat special operations forces.

The MDCI analyst retains copies of the MDCISUM to provide a historical database for future use; to use the preparation of CI threat assessments; and to update the CI estimate. The MDCISUM usually accompanies the graphic intelligence summary prepared by the ACE. This allows commanders to view them simultaneously. The MDCISUM, like the graphic intelligence summary, is an extremely valuable tool. It gives the commander critical information in a concise, graphic manner.

Counterintelligence Threat Assessment:
The CI threat assessment is a four-paragraph statement which is published as often as necessary or when significant changes occur, depending on the situation and the needs of the commander. As a general rule, the CI threat assessment is disseminated by the ACE with every third or fourth MDCISUM. The CI threat assessment provides justification for CI target nominations, a primary goal of CI analysis. Essentially, the CI threat assessment provides the following to the consumer:

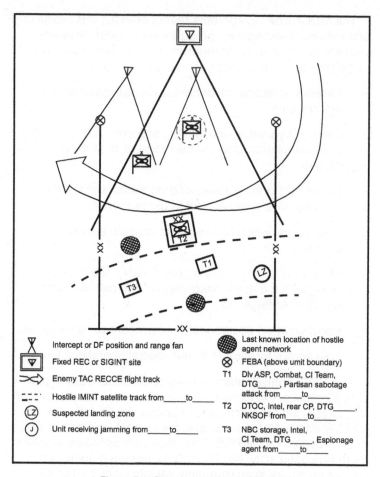

Figure 5.1. Sample graphic MDCISUM

- A quick overview of significant activity during the reporting period.

- An assessment of the intelligence damage.

- A projected assessment of enemy activity for the next reporting period.

- Target nominations.

The CI threat assessment is a valuable means for providing peacetime assessment to commanders, activities, or

64

operations shown in Figure 5-2. This assessment also satisfies the NATO requirement for a CI summary (INTSUM-CI).

CI Situation Overlay:

The CI situation overlay is a composite of the functional area
overlay prepared by the subject matter experts assigned to perform CI analysis. The CI situation overlay incorporates the most
important information from each of the other overlays. The functional area overlay serves as the "working" overlay, while the CI
overlay is the master and serves as the briefing overlay. It should
be ready for briefings at all times. Ordinarily, the senior MDCI
analyst is responsible for maintaining the overlay; however, its
preparation is a collective effort of all members of the CI team.

CI Estimate:

The CI estimate is a composite study containing information
from each functional area pertaining to a specified contingency area. It is a dynamic document prepared during peacetime and refined and updated continuously. The CI estimate
addresses all friendly AOs with the strongest emphasis on
the rear area. The rear operations IPB process is tied to the
development of the CI estimate. Types of information contained in these estimates vary depending on the contingency
area. They generally contain discussions on friendly deployment (including friendly critical nodes) and enemy intelligence
collection capabilities and operations (such as sabotage or
unconventional warfare). The following are examples of information found in an estimate:

- CONUS base.

- Major supply routes.

- Rail lines.

- Points of entry.

- Air and sea lanes.

- Air points of departure and sea points of departure.

- Staging areas.

- Maneuver areas.

- Host nation support and nature of resistance in any US AO.

- Assessment of threats to the logistic system.

- Enemy multidiscipline collection capabilities.
- Level I or II threats.

(CLASSIFICATION)

1. ENEMY ACTIVITY DURING PERIOD ____ TO ____ (LIST DTGs)

a. HUMINT: Summarize in one paragraph all known HUMINT activity during the reporting period. Compile data from HUMINT situation overlay, matrices, link diagrams, and MDCISUMs.

b. SIGINT: Summarize in one paragraph all known SIGINT activity during the reporting period. Compile data from SIGINT situation overlay, matrices, direction charts, and MDCISUMs.

c. IMINT: Summarize in one paragraph all known IMINT activity during the reporting period. Compile data from IMINT situation overlay, matrices, pattern and analysis charts, and MDCISUMs.

d. Other: Summarize all other enemy activity that is not already addressed using the same analytical tools.

2. INTELLIGENCE DAMAGE ASSESSMENT FOR THE PERIOD ____ TO ____ (LIST DTGs)

Briefly assess the intelligence damage to the friendly units for which the assessment is being prepared. Assessment is based on enemy collection activities that were traced, analyzed, and reported in MDCISUMs and were measured against the friendly force operations profile and countermeasures implemented by the friendly force. Coordination with G3 OPSEC staff element is essential in preparing this paragraph.

3. PROJECTED ENEMY ACTIVITY ASSESSMENT FOR THE PERIOD ____ TO ____ (LIST DTGs)

a. HUMINT: Using the same analytical tools identified in paragraph 1A above, plus IPB, project or assess enemy HUMINT activity for the next reporting period.

b. SIGINT: Using the same analytical tools identified in paragraph 1B above, plus IPB, project or assess enemy SIGINT activity for the next reporting period.

---

c. IMINT: Using the same analytical tools identified in paragraph 1C above, plus IPB, project or assess enemy IMIN T activity for the next reporting period.

d. Other: Using the same analytical tools identified in paragraph 1D above, plus IPB, project or assess all other enemy activity for the next reporting period that is not otherwise addressed in the HUMINT, SIGINT, or IMINT assessments.

4. TARGET NOMINATION

a. Exploitation: Using aforementioned information and all other analytical tools, identify any targets worthy of exploitation. Provide recommended time frames, locations, methods of exploitation, justification, and any other pertinent data.

b. Neutralization: Using aforementioned information and all other analytical tools, identify any targets worthy of neutralization. Provide recommended time frames, methods of neutralization, locations, justification, and any other pertinent data.

c. Destruction or Elimination: Using aforementioned information and all other analytical tools, identify any targets worthy of destruction or elimination. Provide recommended methods for engagement, time frames, locations, justification, and any other pertinent data.

NOTE: All target nominations must have G2 or G3 approval before dissemination or presentation to the commander or his designated representative for decision. Coordination with appropriate elements, consistent with type nomination, is essential.

(CLASSIFICATION)

Figure 5-2. CI threat assessment

These considerations necessitate complete CI analysis of the threat and an assessment of friendly critical nodes and targets. Some of these friendly targets will be identified almost out of common sense, but others will require a concerted analytical effort. In preparing the CI estimate, the team should first concentrate on identifying friendly critical nodes and targets and then examine the threat. It should then evaluate the target with respect to their relative criticality, accessibility, vulnerability, and the potential effect of their destruction. This is done for both friendly and enemy targets under the purview of the team. Figure 5-3 is an example of the CI estimate.

(CLASSIFICATION)

Headquarters
Place, Date, and Zone
CI Estimate Number____

References: Maps, charts, or other documents.

1. MISSION: The restated mission determined by the commander.

2. AREA OF OPERATIONS: Summarizes the analysis of the AOs.

  a. Weather.

  (1) Factors. Include light data and either a weather forecast or climatic information, as appropriate. Use appendixes for graphic representation or weather factors and other details.

  (2) Effect on enemy COAs. Discuss the effects of weather on possible enemy COAs (for example, sabotage, subversion, raids, air operations) in terms of mobility, concealment, and logistic sustainability. Discuss in terms of Level I or II threat, and enemy all-source intelligence collection operations.

  b. Terrain.

  (1) *Existing situation.* Use graphic symbols where possible especially for cover, concealment, and logistic sustainability. Use annexes for detailed information. Information covering observation, fields of fire, obstacles, key terrain, and approaches to the probable target aid in determining insurgent terrain. Also, consider graphics for critical facilities and buildings.

  (2) *Effect on enemy COAs.* Discuss in the same manner as for effects of weather in a(2) above. Discuss in detail those areas favorable and unfavorable to the levels of threat I or II enemy all-source intelligence collection opportunities.

---

(3) *Effect on own COAs.* Discuss in the same manner as for effects of weather in a(2) above. Note the positive or detrimental effects on response forces and defensive measures.

c. **Other characteristics.** Include in subparagraphs any of the following factors or characteristics which pertain to friendly area activity (emphasis on rear area): population, ethnicity, religious makeup, literacy rate, medical considerations, economic stability, transportation, mass media, public services, and current political situation.

3. ENEMY SITUATION: Information on the enemy which will permit later development of enemy capabilities and vulnerabilities and refinement of these capabilities into specific COAs and their adoption.

a. **Disposition.** Reference overlays, enemy situation maps, or previously published documents. Location of potential threat forces may be difficult to pinpoint, but the greater the detail, the more thorough the analysis. Separate by level and type of threat (that is, combat Levels I and II threats) or intelligence.

b. **Composition.** Summary of the OB of conventional Levels I and II threats, intelligence collection units and elements, and the structure or organization of paramilitary and/or terrorist groups. Separate by level and type of threat.

c. **Strength**. This listing assists in developing enemy capabilities and vulnerabilities. Conventional and intelligence threats are discussed as in a regular intelligence estimate.Terrorist, paramilitary, and other threats need to be assessed based on support from populace, monetary base, supplies, armament, personnel, and other pertinent considerations. Use subparagraphs to address the different threats. Separate by level and type of threat.

d. **Recent and present significant activities.** Items of information are listed to provide bases for analysis to determine relative probability of adoption of specific COAs and enemy vulnerabilities. Enemy failures to take expected actions are listed as well as positive information. Include recent all-source collection activities, terrorist actions, and other indications.

e. **Peculiarities and weaknesses.** For conventional and intelligence collection threats, discuss as in an intelligence estimate. For terrorist, paramilitary, or other unconventional threats, discuss such pertinent information as leadership (key personalities), equipment, finances, and relations with the local populace.

4. ENEMY CAPABILITIES: Based on all the previous information and analysis, develop and list enemy capabilities to conduct operations against the friendly area, with emphasis on the rear area. The listing provides a basis for analyzing the available information to arrive at those capabilities that the enemy can adopt as specific COAs and their relative probability of adoption. Separate items by levels and type of threat.

a. **Enumeration.** State what, when, where, in what strength, and by what method for each threat.

b. **Analysis and discussion.** Each enumerated threat is discussed in terms of indicators of adoption or rejection. The intent is to assess the likelihood of a given threat taking a given action. Consider all information previously recorded in this estimate. Some threats may not have any indicators of rejection listed due to insufficient data.

5. POTENTIAL ENEMY TARGETS: Based on all previous information and analysis, develop, to the extent possible, a listing of potential enemy targets. Ensure you can identify, at a minimum: target identity, capability, location or projected location, and projected intentions. Ascertain if targets can be exploited, neutralized, destroyed, or eliminated. Use subparagraphs and/or annexes as needed.

6. CONCLUSIONS:

a. **Effects of AOs on own COAs.** Indicate weaknesses in ability of response forces to react in defensive measures.

b. **HVT analysis based on the criticality, accessibility, recuperability, vulnerability, and effect (CARVE) format.** Such targets range from bridges to friendly units, public services, and key facilities. Complete for both friendly and enemy targets.

c. **Probable enemy COAs.** COAs are listed in order of relative probability of action. However, insufficient data may only permit the probable level of threat for a given target.

d. **Enemy vulnerabilities.** List the effects of peculiarities and weaknesses that result in vulnerabilities that are exploitable. Annexes (as required): Annexes may include graphic analysis products which support the estimate such as link diagrams, association matrices, rear operations IPB products, or black, white, and gray lists distributed to units requiring them.

(CLASSIFICATION)

Figure 5-3. CI Estimate

Appendix A
Counter-Human Intelligence Techniques and Procedures

General

Appendix A contains information on operations of CI interest, and on the C-HUMINT analysis performed by MDCI analysts. It contains basic information for the C-HUMINT agent and analyst as well as the interrogator. The appendix describes those procedures employed to conduct two types of investigations as well as the legal principles important to successful completion of investigations. The investigative techniques and legal principles are presented to help expedite investigations and keep them from being bogged down and rendered ineffective by technical and legal errors. These general principles are reinforced by investigative SOPs promulgated by those commands that conduct investigations.

Contents

C-HUMINT, to include investigations, operations, collections, and analysis and production, have their own unique techniques and procedures. These techniques and procedures include

- Basic Investigative Techniques and Procedures.

- Investigative Legal Principles.

- Technical Investigative Techniques.

- Screening, Cordon, and Search Operations.

- Personalities, Organizations, and Installations List.

- Counter-Human Intelligence Analysis.

- Personnel Security Investigations.

- Counterintelligence Investigations.

Each is covered in some detail in this appendix.

73

Section I
Basic Investigative Techniques
and Procedures
to
Appendix A
Counter-Human Intelligence Techniques
and Procedures

A-I-1. General.

The basic investigative techniques and procedures described in this section apply to both primary types of investigations: PSI and CI (also called SAEDA) investigations. Specific information for PSI is contained in Section VII to this appendix, while CI investigations is contained in Section VIII to this appendix.

A-I-2. Legal Principles.

a. Most CI investigations go beyond arrest and prosecution of suspects. If an investigation cannot evolve into a more productive CI operation, and when further exploitation is not possible, the objectives must be deterrence or prevention and prosecution of the suspects. Therefore, the procedures used during an investigation must be compatible with the requirements for prosecution.

b. Investigations must be conducted in accordance with the principles of law and the rules of evidence which govern the prosecution of any criminal activity. CI personnel must have a thorough understanding of the legal principles (see Section II to Appendix A) and procedures involved in conducting an investigation for three reasons:

    (1) To strictly apply them in all investigative activity.

    (2) To be guided by them in cases not foreseen, when there is no time to seek specific guidance or assistance.

(3) To be able to recognize those cases where specific guidance or assistance and approval must be
obtained before proceeding further.

c. Basic legal principles will always apply to CI investigative situations. The legal principles are designed to
ensure that the legal rights of subjects or suspects are
observed. It is important to ensure that the potential
ability to prosecute any given case is not jeopardized
by illegal or improper CI investigative techniques. In
addition, CI personnel involved in investigative activities must obtain advice from the Staff Judge Advocate
(SJA) or legal officer to implement recent court decisions interpreting statutes and regulations.

d. In cases where prosecution is a possibility, CI investigative
personnel should brief the SJA trial counsel in the initial
stages of the investigation. After coordination with the SCO
and obtaining command approval, it will support the prosecution's case and provide insight to the CI agent regarding
case direction. AR 195-5 and FM 19-20 cover the legal
aspects of gathering, handling, and controlling evidence.

A-I-3. Investigative Techniques.

Checking files and records for pertinent information on the
subject of the investigation is the first action in CI investigations. Checks should begin with local unit files and expand
to include the Investigative Records Repository and other
military services and civilian agencies. (No Army element
will retain in its files any information which is prohibited by
AR 381-10.) The full exploitation of records examination as
an investigative tool depends on several factors which the CI
agent must consider.

a. The CI agent must know what, where, by whom, and
for what purpose records are maintained throughout
the AO. Upon assignment to an operational unit, the
initial orientation should stress that the agent be thoroughly familiar with records that may be of assistance
in investigations.

b. Most records are available to the CI agent upon official
request. If all efforts to obtain the desired information

through official channels are unsuccessful, the information or records cannot be subpoenaed unless legal proceedings are initiated.

c. There are occasions when documentary information or evidence is best obtained through other investigative means. The possibility of intentional deception or false information in both official and unofficial records must always be considered. Because data is recorded in some documentary form does not in itself ensure reliability. Many recorded statistics are untrue or incorrect, particularly items of biographical data. They are often repetitious or unsubstantiated information provided by the SUBJECT himself and are not to be confused with fact.

d. Reliability of records varies considerably according to the area and the status of the agency or organization keeping the records. Records found in highly industrialized areas, for example, are more extensive and generally far more reliable than those found in underdeveloped areas. Until experience with a certain type of record has been sufficient to make a thorough evaluation, treat the information with skepticism.

e. If the record is to be used in a court or board proceeding, the manner in which it is copied, extracted, or preserved will have a bearing on its use as evidence.

f. In CI investigations, the absence of a record is often just as important as its existence. This is especially important in the investigation of biographical data furnished by the SUBJECT of a CI investigation. The systematic and meticulous examination of records to confirm or refute a SUBJECT's story is very often the best means of breaking the cover story of an enemy intelligence agent.

g. The types and content of records vary markedly with the AO. Regardless of the area, the CI agent must be aware of the types of records the agent may use in conducting investigations. Available records include police and security agencies, allied agencies, vital statistics, residence registration, education, employment, citizenship, travel, military service, foreign military records, finance records, and organization affiliation.

(1) Police and security agencies. Some major types of records which are often of value are local, regional, and national police agencies. Most nations maintain extensive personality files covering criminals, SUBJECTs, victims, and other persons who have come to official police attention because of actual or alleged criminal activity. Police interest in precise descriptive details, including photographs and fingerprint cards, often make police records particularly valuable and usually more reliable than comparable records of other agencies. Police and security agency files are usually divided into subcategories. The CI agent must be familiar with the records system to ensure all pertinent files actually have been checked.

(2) Allied agencies. Access to records of allied intelligence agencies often depends on the personal relationship between the CI representative and the custodian of the records of interest. Such examinations are normally the assigned responsibility of an LNO. Liaison also may be necessary with other agencies when the volume of records examinations dictate the need for a single representative of the CI element. At times it may be necessary, due to the sensitivity of a particular investigation, to conceal specific interest in a person whose name is to be checked. In this instance, the name of the individual may be submitted routinely in the midst of a lengthy list of persons (maybe five to seven) who are to be checked.

(3) Vital statistics. The recording of births, deaths, and marriages is mandatory in nearly every nation, either by national or local law. In newly developed countries, however, this information may be maintained only in family journals, bibles, or in very old records. In any case, confirmation of such dates may be important. The records sought may be filed at the local level, as is usually the case in overseas areas; or they may be kept at the state or regional level, such as with state bureaus of vital statistics in the US. Rarely will original vital statistics

records on individuals be maintained centrally with a national agency.

(4) Residence registration. Some form of official residency registration is required in most nations of the world. The residence record may be for tax purposes, in which case it probably will be found on file at some local fiscal or treasury office. When the residence record is needed for police and security purposes, it is usually kept in a separate police file. Residence directories, telephone books, and utility company records also may be used.

(5) Education. Both public and private schools at all levels, from primary grades through universities, have records which can serve to verify background information. The school yearbook or comparable publication at most schools usually contains a photograph and brief resume of the activities of each graduating class member. These books are a valuable record for verification and as an aid to locating leads. Registrar records normally contain a limited amount of biographical data but a detailed account of academic activities.

(6) Employment. Personnel records usually contain information on dates of employment, positions held, salary, efficiency, reason for leaving, attendance record, special skills, and biographical and identifying data. Access to these records for CI agents is relatively simple in the US but may prove difficult in some overseas areas. In such areas, it may be possible to obtain the records through liaison with local civil authorities or through private credit and business rating firms. Depending on the AO, there may be either local, regional, or national unemployment and social security program offices. Records of these offices often contain extensive background material. In most cases, these data represent unsubstantiated information provided by the applicant and cannot be regarded as confirmation of other data obtained from the same individual. Records of the US Social Security Administration can be obtained only by the

Department of Justice through written request in cases involving high-level security investigations.

(7) Citizenship. Immigration, nationalization, passport, and similar records of all nations contain data regarding citizenship status. In most instances, an investigation has been undertaken to verify back-ground information contained in such records; therefore, these records are generally more reliable than other types. The records of both official and private refugee welfare and assistance agencies also provide extensive details relating to the citizen-ship status of persons of CI interest. As a general rule, refugee records (particularly those of private welfare groups) are used as a source of leads rather than for verification of factual data, since they have been found to be unreliable in nearly all AOs.

(8) Travel. A system of access to records of interna-tional travel is especially important to CI operations in overseas areas. Such records include customs records, passport and visa applications, passen-ger manifests of commercial carriers, currency exchange files, transient residence registrations, private and government travel agency records, and frontier control agency files. The State Department maintains passport information on US citizens; this information is available by means of the NAC. Additionally, some units maintain records of all personal foreign travel by any assigned member.

(9) Military service. Records of current and past mem-bers of the armed services of most nations are detailed and usually accurate.

(a) CI agents will encounter no difficulty in obtain-ing access to US military service records on official request. If a service member changes branches, has a break in service, or is hospital-ized, certain elements of information must be furnished to the control office so these records can be located for review, if necessary. For those personnel who have changed branches of service, the control office will need the individual's social security number (SSN), full name, and date and place of birth. An individual's field 201 file is retired when a break in service occurs. To obtain it for review, the control office needs the individual's full name and any former service numbers.

(b) The Adjutant General's offices or control branch files may be more complete than the individual's field 201 file, particularly if the individual has had National Guard or Reserve duty as a commissioned or warrant officer.

(c) Retired Army hospital records are filed by the hospital name and year. Consequently, the name of the hospital and the correct year are required if a search of hospital records is necessary.

(10) Foreign military records. Access to foreign military records in overseas areas may be difficult. In cases where it is not possible to examine official records, leads or pertinent information may be obtained from unofficial unit histories, commercially published documents, and files of various veterans organizations. Since listing or claiming military service is a convenient means of accounting for periods of time spent in intelligence activities or periods of imprisonment, it is frequently a critical item in dealing with possible agents of a FIS. Special effort must be made to locate some form of record which either confirms or denies an individual's service in a particular unit or the existence of the unit at the time and place the individual claims to have served. OB and personality files of various intelligence services also may be helpful.

(11) Finance records. Finance records are an important source of information. They may provide information to indicate whether a person is living beyond one's means. They may provide numerous leads such as leave periods and places, and identification of civilian financial institutions.

(12) Organization affiliation. Many organizations maintain records which may be of value to a particular investigation. Examples are labor unions; social,

scientific, and sports groups; and cultural and sub-versive organizations. CI agents should research these organizations. But when seeking sources of information, the CI agent must be thoroughly familiar with the organization before attempting to exploit it. Organizations are often established as front groups or cover vehicles for foreign intelligence operations.

h. Having determined which records may include informa-tion pertinent to an investigation, the CI agent must select the best means to gain access and examine or copy them.

(1) The following are the procedures a CI agent should follow to gain access to records:

(a) Contact the records custodian to include medi-cal records custodian.

(b) Use proper credentials to establish identity as a US Army Special Agent.

(c) State the purpose of the inquiry.

(d) Ask for any available information.

(2) The above procedures are commonly used in PSIs, but may also be used in certain phases of CI investigations.

(3) The CI agent may conduct local agency checks by mail or telephone when time, money, or physi-cal constraints prevent personal contact with the local agency records custodian. This procedure is discouraged unless personal contact is impossible. However, before an arrangement of this nature begins, coordination must be made with higher headquarters and the local CI agent in charge. Liaison is reciprocal cooperation between an agency with records of interest and the unit. It may include authorization for a liaison representative to conduct records checks on an exchange basis within limitations imposed by higher headquarters. This type of liaison is normally the responsibility of a designated LNO or an additional duty for a CI agent when discreet checks are not required.

81

(4) There is a risk factor with records checks. Exposure of the SUBJECT's name and the fact that he is under investigation may alert the SUBJECT. One way to obtain record information is to include the SUBJECT's name in a list of persons whose records are to be checked, thus pointing no spotlight at the SUBJECT.

A-I-4. Interview Techniques.

The interview is a structured conversation designed to obtain information from another person who is known or believed to possess information of value to an investigation. It is an official encounter, and, as such, must be conducted in accordance with the rules of evidence and other legal principles. Persons mentioned during interviews are all potential sources of information. Therefore, the CI agent should attempt to influence this person in a positive way so he will want to provide the needed information. Establish rapport between the CI agent and the interviewee; use the proper interview techniques and the basic tools of human communication. The interview can be categorized into one of several types:

a. PSI Reference Interview.

b. PSI SUBJECT Interview.

c. CI (SAEDA) Walk-in Interview.

d. CI (SAEDA) Source Interview.

e. CI (SAEDA) SUBJECT Interview.

NOTE: PSI interviews are discussed in detail in Section VII to Appendix A; CI interviews are discussed in detail in Section VIII to Appendix A.

A-I-5. Interrogation Techniques.

Interrogation is obtaining the maximum amount of usable information through formal and systematic questioning of an individual. Apply the principles and techniques of interrogation contained in FM 34-52 to CI interrogations. CI interrogations should be conducted by at least two CI agents.

a. The CI agent uses interrogation techniques when encountering a hostile source or SUBJECT. The self-preservation instinct is stimulated in an individual who is considered a SUBJECT. This deep-rooted reaction is frequently reflected in stubborn resistance to interrogation. The SUBJECT may consider the interrogation as a battle of wits where the SUBJECT has much to lose. The SUBJECT may look upon the CI agent as a prosecutor.

b. When interrogating a SUBJECT, the CI agent must keep in mind the two-fold objective of the interrogation:

(1) Detection and prevention of activity that threatens the security of the US Army.

(2) Collection of information of intelligence interest.

c. Generally, the CI agent works toward obtaining intelligence information from the SUBJECT, leading to a confession admissible in court.

d. When preparing for an interrogation, the CI agent should—

(1) Gather and digest (complete familiarization) all available material concerning the SUBJECT and the case.

(2) Be familiar with those legal principles and procedures which may apply to the case at hand. Legal requirements may differ depending on—

(a) Whether the US is at war or in a military occupation.

(b) SOFAs.

(c) Whether the SUBJECT is a US citizen or a member of the US Armed Forces.

(d) Whether the SUBJECT is an EPW.

(3) Determine the best way to approach the SUBJECT. Previous investigative efforts may have determined that the SUBJECT is under great psychological pressure; therefore, a friendly approach might work best. The CI agent should carefully consider the

approach and the succeeding tactics, to ensure that nothing the agent does will cause the SUBJECT to confess to a crime he or she did not commit.

e. Before an interrogation, the CI agent must ensure the following:

(1) The interrogation room is available and free of distractions.

(2) If recording equipment is to be used, it is installed and operationally checked.

(3) All participants in the interrogation team are thoroughly briefed on the case and interrogation plan.

(4) Sources or other persons to be used to confront the SUBJECT are available.

(5) Arrangements are made to minimize unplanned interruptions.

(6) As appropriate, arrangements are made for the SUBJECT to be held in custody or provided billeting accommodations.

f. When conducting the interrogation, apply the basic techniques and procedures outlined in FM 34-52. The following points are important:

(1) Use background questioning to provide an opportunity to study the SUBJECT face-to-face.

(2) Avoid misinterpretation and impulsive conclusions. The fact that the person is suspected may in itself create reactions of nervousness and emotion.

(3) Do not allow note-taking to interfere with observing the SUBJECT's reaction.

(4) Seek out all details concerning the SUBJECT's implication in a prohibited activity.

(5) Examine each of the SUBJECT's statements for its plausibility, relationship to other statements or to known facts, and factual completeness. Discrepancies which require adjustment frequently weaken the SUBJECT's position.

(6) Attempt to uncover flaws in details not considered relevant to the issue; finding the story's weakness is the key to a successful interrogation.

(7) Build up to a planned final appeal as a sustained and convincing attack on the SUBJECT's wall of resistance. Eloquent and persuasive reasoning and presenting the facts of the case may succeed where piecemeal consideration of evidence failed to produce a confession. This appeal may be based on overwhelming evidence, on contradictions, story discrepancies, or the SUBJECT's emotional weaknesses.

(8) Obtain a sworn statement if the SUBJECT wants to confess. If the SUBJECT has been given an explanation of individual rights under Article 31, Uniform Code of Military Justice (UCMJ), or the 5th Amendment to the US Constitution, any unsworn statement normally can be used in court. If the SUBJECT is neither a US citizen nor a member of the armed forces, requirements will be stipulated in the unit's SOP.

g. CI agents may use polygraph examinations as an aid to CI interrogations and investigations of intelligence operations, but only at the direction of higher headquarters.

A-I-6. Elicitation.

Elicitation is gaining information through direct communication, where one or more of the involved parties is not aware of the specific purpose of the conversation. Elicitation is a planned, systematic process requiring careful preparation.

a. Preparation. Always apply elicitation with a specific purpose in mind.

(1) The objective, or information desired, is the key factor in determining the SUBJECT, the elicitor, and the setting.

(2) Once the SUBJECT has been selected because of his or her access to or knowledge of the desired information, numerous areas of social and official dealings may provide the setting.

(3) Before the approach, review all available intelligence files and records, personality dossiers, and knowledge possessed by others who have previously dealt with the SUBJECT. This will help to determine the SUBJECT's background, motivation, emotions, and psychological nature.

b. Approach. Approach the SUBJECT in normal surroundings to avoid suspicion. There are two basic elicitation approaches: flattery and provocation. The following variations to these approaches may be used:

(1) By appealing to the ego, self-esteem, or prominence of the SUBJECT, you may be able to guide him or her into a conversation on the area of operation.

(2) By soliciting the SUBJECT's opinion and by insinuating that he or she is an authority on a particular topic.

(3) By adopting an unbelieving attitude, you may be able to cause the SUBJECT to explain in detail or to answer out of irritation. The CI agent should not provoke the subject to the point where rapport is broken.

(4) By inserting bits of factual information on a particular topic, you may be able to influence the SUBJECT to confirm and further expound on the topic. Use this approach carefully since it does not lend itself to sudden impulse. Careless or overuse MW of this technique may give away more information than gained.

(5) By offering sincere and valid assistance, you may be able to determine the SUBJECT's specific area of interest.

c. Conversation. Once the approach has succeeded in opening the conversation, devise techniques to channel the conversation to the area of interest. Some common techniques include—

(1) An attempt to obtain more information by a vague, incomplete, or a general response.

(2) A request for additional information where the SUBJECT's response is unclear; for example, "I agree; however, what did you mean by...?"

(3) A hypothetical situation which can be associated with a thought or idea expressed by the SUBJECT. Many people who would make no comment concerning an actual situation will express an opinion on hypothetical situations.

A-I-7. Opposite Sex Interview.

During the preliminary planning for an interview of a member of the opposite sex, the CI agent must place emphasis on avoiding a compromising situation. This is particularly true when the person to be questioned is a SUBJECT or is personally involved in a controversial matter.

a. Embarrassment is inherent in any situation where a member of the opposite sex questions a SUBJECT concerning intimate, personal matters. The CI agent must make provisions to have present a member of the same sex as SUBJECT when the subject matter or questions might prove embarrassing to the SUBJECT.

b. Before interrogating a member of the opposite sex, advise the individual about the right to request the presence of a person of the same sex. As an alternative, a CI agent of the same sex could conduct the interview.

c. In any event, the CI agent should ensure that a third person is present, or within constant hearing distance, during any interview of a member of the opposite sex. This person must possess the necessary security clearance for the subject matter to be discussed.

d. Should questions arise during the interview that could prove embarrassing, the CI agent, before asking such questions, should advise the individual being questioned that such questions will be asked.

e. The CI agent may have another person present during such interviews, even though the Source or SUBJECT

does not make a request. If the individual being questioned objects to the presence of another individual
and would be less cooperative in another person's
presence, have that objection and its basis reduced
to writing and signed by the Source or SUBJECT, and
then have the other person in attendance withdraw. If
the objection is merely to the visual presence of the
third party and not to the third party listening to the
statement, make provisions to have the other person in
attendance within normal voice range of the place of
questioning, but out of sight.

f. In those investigations where the member of the opposite
sex has a recent history of serious mental or nervous
disorders, another member of the opposite sex must be
present during the interview.

Section II
Investigative Legal Principles
to
Appendix A
Counter-Human Intelligence Techniques
and Procedures

A-II-1. General.

This section looks at the legal basis for CI activities. It begins with a short discussion of the EO and the implementing of DOD and Army regulations covering intelligence activities. Section II provides strict guidelines and procedures requiring a thorough knowledge of criminal law, methods of obtaining and processing evidence, an individual's rights, and regulation oversight. It lays the groundwork MW for the investigative and reporting sections to follow.

a. AR 381-10 sets policies and procedures governing the conduct of intelligence activities by Army intelligence components. It proscribes certain types of activities. AR 381-20 implements EO 12333, proscribing certain types of activities and strictly regulating actions commonly referred to as special collection techniques. Currently, it is the governing regulation applicable to MI assets. Neither regulation is a mission statement nor does it task entities or individuals to collect information. Instead, they constitute rules of engagement for collecting information on US persons.

b. The importance of understanding these regulations and following its guidelines to accomplish the mission cannot be overstressed. It is imperative that the individuals requiring the collection of information read and understand the—

    (1) Restrictions placed on collecting information concerning US persons.

    (2) Definition of the terms "collection" and "US person."

    (3) Retention and dissemination of that information.

(4) Use of special collection techniques.

(5) Process of identifying and reporting questionable activities or activities in violation of these regulations, US law, or the US Constitution.

c. CI agents should seek constant assistance from their local judge advocate on the interpretation and application of the procedural guidelines contained in these regulations. Rely only on a trained lawyer's interpretation when seeking to implement any of the special collection procedures.

d. Because of the nature of this work, CI agents must at least understand the basic legal principles. Decisions made by the CI agent are frequently guided by legal concepts. Only a CI agent who is familiar with governing legal principles is able to conduct these tasks efficiently, and within the parameters of the law.

A-II-2. Criminal Law.

Although the Army agent's criminal investigatory responsibility is limited to crimes involving national security, the CI agent must understand general criminal law as well.

a. A crime almost always requires proof of a physical act, a mental state, and the concurrence of the act and the mental state. Criminal law is not designed to prosecute persons who commit acts without a "guilty" intent. For example, a reflex action by an epileptic would not subject the epileptic to criminal liability. Additionally, criminal law is not directed at punishing individuals who privately plan criminal activity when those plans are not combined with action.

b. The act required for criminal prosecution may be a willful act or an omission to act when so required by law. Espionage, the unauthorized receipt of classified information, is an act punishable as a crime. The failure to report missing classified documents is an example of an omission to act punishable as a crime. The law recognizes several different mental states. National security crimes frequently distinguish between specific and general intent. Specific intent requires that an individual act with one of two mental elements:

(1) Purposely, indicating a desire to cause a particular result.

(2) Knowingly, indicating that the individual is substantially certain that the act will cause the result.

c. General intent is indicated by an individual who commits an act with knowledge that an unjustifiable risk of harm will occur. This definition of general intent, and the example below, are specific to the crime of sabotage (Title 18, US Code, Sections 2151-2156) which contains both specific and general elements, and are not necessarily representative of the notion of general intent as a generic criminal law concept.

Alpha, a service member, places a bolt in the engine of an F-16 aircraft. If Alpha placed the bolt in the engine with the specific intent to harm the national defense by the loss of a combat aircraft, Alpha may be convicted of sabotage. However, if Alpha placed the bolt in the engine because of a dissatisfaction with the service, and not to harm specifically the national defense, he may still be convicted of sabotage. This is true because Alpha acted in a manner indicating that he knew that an unjustifiable risk of harm would result from the act of tampering with the aircraft. Since the loss of the aircraft would hamper the national defense effort, Alpha generally intended to commit sabotage in knowing of the risk to the national defense by the act, and yet, consciously disregarding the risk.

d. The inchoate or incomplete crimes are of particular importance to CI agents. These crimes are conspiracy, attempt to commit a specific crime, and solicitation to commit a crime.

(1) Article 81, United States Manual for Courts-Martial, 1984, identifies the following elements of conspiracy:

(a) That the accused entered into an agreement with one or more persons to commit an offense under the code.

(b) That while the agreement continued to exist, and while the accused remained a party to the agreement, the accused or at least one of the co-conspirators performed an act for the purpose of bringing about the object of the conspiracy. The overt act required for a conspiracy may be legal:

Alpha and Bravo conspire to commit espionage. Bravo rents a post office box in which classified information is to be placed to distribute to a foreign agent. At the time Bravo rents the box and before any classified material has been placed in the box, Alpha and Bravo may be arrested and charged with conspiracy to commit espionage.

(2) Article 80, United States Manual for Courts-Martial, 1984, lists the following elements of attempt:

(a) That the accused did a certain overt act.

(b) That the act was done with the specific intent to commit a certain offense under the code.

(c) That the act amounted to more than mere preparation.

(d) That the act apparently tended to effect the commission of the intended offense:

Alpha intends to commit espionage by receiving classified information from Bravo. Unknown to Alpha, Bravo is actually working for CI. Bravo provides Alpha with blank papers. Alpha is arrested as he picks up the papers. Alpha may be charged with and convicted of attempted espionage.

(3) Article 82, United States Manual for Courts-Martial, 1984, identifies the following elements of solicitation: That the accused solicited or advised a certain person or persons to commit any of the four offenses named in Article 82; and that the accused did so with the intent that the offense actually be committed.

(a) Sedition and mutiny are identified in Article 94, United States Manual for Courts-Martial, 1984, as two of the four offenses that may be solicited. CI agents must, therefore, investigate to determine whether a solicitation to commit an act of sedition or mutiny has occurred.

(b) Solicitation does not require that the substantive crime (sedition) actually be committed. The accused need only advise or encourage another to commit the substantive crime. The remaining two crimes under Article 82 are not of CI interest.

A-II-3. Evidence.

a. The United States Manual for Courts-Martial, 1984, contains the rules of evidence applicable in courts-martial. AR 15-6 governs most administrative proceedings. AR 15-6 limits the admissibility of evidence to it being relevant and material. The CI agent must only ask whether the particular piece of evidence tends to prove or disprove a fact of consequence in the adjudication. If it does, it is admissible in an administrative hearing. The only exception to this rule is if the evidence violates the severely limited exclusionary rules applicable to administrative hearings.

b. A respondent's confession or admission, obtained by unlawful coercion or inducement likely to affect its truthfulness, will not be accepted as evidence against that respondent in an administrative proceeding. The failure to advise a respondent of Article 31, UCMJ, or Fifth Amendment rights does not, by itself, render the confession or admission inadmissible in an administrative hearing. Evidence found as the result of a search, conducted or directed by a member of the armed forces acting in an official capacity who knew the search was illegal, will not be admissible in an administrative hearing.

c. Evidence unlawfully obtained through any of the ways covered in this appendix is generally inadmissible as evidence against the suspect or accused. Any other evidence subsequently obtained or derived as a result of this evidence is likewise inadmissible. Whether the use of a particular item in evidence violates an individual's rights is usually a complex and technical determination. The CI agent should obtain advice from the local SJA specified by the unit in its SOP.

A-II-4. Rights.

The remainder of this section discusses the constitutional rights of a suspect or accused person during an investigation and the legal issues involved when obtaining evidence by means of interviews or interrogations, searches and seizures, and apprehensions. A violation of a person's legal rights will substantially, if not completely, reduce the chances of a successful criminal prosecution.

a. Interviews and interrogations. Article 31, UCMJ, and the Fifth Amendment to the US Constitution prohibit government agents from compelling any person to incriminate oneself or to answer any question, the answer to which may tend to incriminate. One may remain absolutely silent and not answer any questions. To be admissible against a person, a confession or admission must be voluntary. A statement obtained through coercion (or other unlawful inducement) is termed involuntary. Physical violence, confinement, and interrogation to the point of exhaustion are examples of acts which may produce involuntary statements.

(1) Courts will use a "reasonable person" test to determine whether the investigator should have considered the individual a suspect and, therefore, given an advisement of rights under Article 31, UCMJ. If individuals appear confused as to their rights or their status, the CI agent should make every reasonable effort to remove the confusion. As the factors that affect a proper warning may be changed by court decision, the CI agent should seek appropriate advice on a continuing basis from a judge advocate. Under Article 31, UCMJ, anyone subject to the UCMJ who is suspected of a crime and who is interviewed by an agent who is also subject to the UCMJ must be advised of his rights.

(2) To enforce the constitutional prohibition against psychologically coerced confessions, Congress, the Supreme Court, and the Court of Criminal Appeals (formerly Court of Military Appeals) have acted to require government agents to advise all suspects and accused persons of their legal rights before questioning. Failure to give the advisement, even to a suspect who is a lawyer or CI agent, will result in the exclusion of the interviewee's statements at trial. Therefore, before beginning suspect interviews, the CI agent informs the individual of his official position, the fact that the individual is a suspect or accused, and the nature of the offense of which he is accused or suspected. If unsure of the precise charge, the CI agent will explain, as specifically as possible, the nature of the facts and circumstances which have resulted in the individual's being considered a suspect.

(3) If a suspect waives the Article 31, UCMJ, rights, the government must be prepared at trial to prove that the defendant understood the rights and chose to waive them voluntarily, knowingly, and intelligently. This is a greater burden than merely showing that the suspect was read the rights and did not attempt to assert or invoke them. Whether the government can sustain this burden at trial depends largely on the testimony and written record furnished by the CI agent who conducted the interview.

(4) The explanation of rights set forth below replaces all previous explanations of legal rights, including the customary reading of Article 31, UCMJ, and the Fifth Amendment to the US Constitution. A mere recitation of this advisement, however, does not assure that subsequent statements by the suspect will be admissible in court, as it must be shown that the suspect, in fact, understood the rights.

(5) The explanation of rights will not suffice if delivered in an offhand or ambiguous manner. The tone of the interrogator's voice should not suggest

that the advisement is a meaningless formality. It also would be improper for the interrogator to play down the seriousness of the investigation or play up the benefits of cooperating. In short, the interrogator must not, by words, actions, or tone of voice, attempt to induce the individual to waive the right to remain silent or the right to counsel. Such action will be denounced by the courts as contrary to the purpose of the explanation of rights requirement.

(6) The same result will occur if the interrogator accidentally misstates or confuses the provisions of the advisement. In addition, all other evidence offered will become suspect, for besides excluding illegally obtained statements, the court may reject—and administrative boards are free to reject—any evidence which is not the logical product of legally obtained statements.

(7) Tricking, deceiving, or emphasizing the benefits of cooperating with the government have not been declared illegal per se. The methods used must merely be directed toward obtaining a voluntary and trustworthy statement and not toward corrupting an otherwise proper Article 31, UCMJ, and the Fifth Amendment rights advisement. The CI agent administers the advisement of rights, in accordance with DA Form 3881. When the suspect appears confused or in doubt, the interrogator should give any further explanations by way of a readvisement of the rights.

(8) If the interviewee indicates a desire to consult with counsel for any reason, the interrogator should make no further attempt to continue the questioning until the suspect has conferred with counsel or has been afforded the opportunity to do so. The interrogator may not subsequently continue the questioning unless the interviewee's counsel is present (or unless the interviewee voluntarily initiates further contact with the interrogator). If suspects decide to waive any right, such as the right to have counsel present at the interrogation

or the right to remain silent, the interrogator). will inform them that they may reassert their rights at any time.

(9) Under no circumstances will the suspect be questioned until the interrogator is satisfied that the individual understands the rights. The interrogator will ask the suspect to sign a waiver certificate (Part 1, DA Form 3881).

(10) If a military member, subject to the code, is suspected of an offense under the UCMJ, the person is entitled to be represented by an attorney at government expense. This can be a military lawyer of the member's choice or, if the requested lawyer is not reasonably available, a detailed military lawyer from the local Trial Defense Service Office. The suspect may also retain a civilian lawyer at no expense to the government. If an appointed counsel is refused, the suspect must have a reasonable basis for that refusal, for example, obvious incompetency. However, the suspect may not arbitrarily declare the counsel unacceptable.

(11) If the suspect requests a specific civilian lawyer, the interrogator must permit the suspect to retain one, and must not continue the interrogation unless the suspect's lawyer is present when questioning resumes, or the suspect voluntarily initiates the resumption of questioning and declines the presence of counsel. The interrogator should assist the individual in obtaining acceptable counsel. The interrogator may not limit the suspect to one telephone call or otherwise interfere in the assertion of the right to counsel.

(12) Civilians generally are not entitled to have counsel provided for them by the armed services. If a civilian suspect demands an attorney, the interrogator must permit the suspect to retain counsel. If the suspect has no lawyer, the interrogator should aid in obtaining legal counsel by providing the suspect with the names and addresses of local agencies that provide legal services.

(13) Such organizations as Legal Aid and the Lawyer's Referral Service are generally listed in local telephone directories. It is to the interrogator's advantage to aid the suspect, since the interrogator may initiate further interrogation only when the suspect is properly represented and counsel is present at the subsequent interrogation. The interrogator should direct all questions about legal representation to the local SJA.

(14) The interrogator should be prepared to question the SUBJECT about each right. Whenever possible, the interrogator should make a verbatim recording of these questions and answers. If this is not possible, the interrogator should ask the SUBJECT to acknowledge both the advisement of rights and an understanding of each right in writing.

(15) It is highly desirable to obtain oral and written acknowledgments. With evidence of both oral and written acknowledgments, the interrogator is well prepared to rebut any charge that the SUBJECT did not understand the rights.

(16) The interrogator should obtain evidence in writing that the SUBJECT made a conscious and knowledgeable decision to answer questions without a lawyer (or to speak with the assistance of a lawyer). See Figure A-II-1 for questions to ask a SUBJECT to assure an understanding of rights.

(17) If, at any time and for any reason, the SUBJECT indicates in any manner a reluctance to answer any more questions or wants to see a lawyer, the interrogator must stop immediately. The interrogator should make no attempt to persuade the SUBJECT to change his or her mind. If the SUBJECT does not want to stop the interrogation entirely, but chooses to refuse to answer some questions while answering others, the interrogator is under no obligation to continue but should certainly do so in most cases.

- Do you understand that you have the right to have a lawyer of your choice during this interview to advise and assist you?

- Do you also understand that your right to a military counsel means a professional lawyer and not just an officer or military superior? (Military only)

- Do you understand that the Army will provide you with a military lawyer free of charge?

- Do you understand that if you decide to answer questions that you may stop whenever you choose?

- Do you understand that anything you say will be made a matter of written record and can be used against you in a court of law?

Figure A-II-1. Questions to ask a SUBJECT to assure an understanding of the rights.

(18) However, the interrogator must not end the interrogation in a manner calculated to intimidate, induce, or trick the SUBJECT into answering questions that the SUBJECT does not care to answer. Under no circumstances should the interrogator ask the SUBJECT why he or she decided to reassert the rights.

(19) Article 31, UCMJ, also prohibits any use of coercion, unlawful influence, or unlawful inducement in obtaining any other evidence from a SUBJECT or accused. One general rule is that SUBJECTs may not be compelled to provide incriminating evidence against themselves through the exercise of their own mental faculties, as for example by making an oral or written statement. Conversely, a SUBJECT may be compelled to provide evidence if the form of evidence will not be affected by conscious thought, and provided the means of coercion (or compelled production) fall within the limits of fundamental decency and fairness. As examples,

Article 31, UCMJ, does not prohibit the taking of the following nontestimonial evidence from SUBJECTs: fingerprints, blood samples, and hand-writing or voice exemplars. These rules are often quite difficult to apply unless guidance is obtained from a legal officer or judge advocate.

(20) If a SUBJECT has been interrogated without a proper rights advisement, it is possible to correct the defect and proceed after a valid advisement of rights. If the CI agent does not know whether a prior statement was properly obtained, or does know that an impropriety occurred, the agent should provide an additional, accurate advisal of rights, and advise the SUBJECT that any statements given pursuant to defective procedures will not be admissible. This technique will minimize or eliminate the taint of the earlier errors and increase the likelihood of admissibility for the subsequent statement. (Rule 304, Manual for Courts-Martial, 1984.)

b. Search and seizure. The Fourth Amendment of the US Constitution, the Manual for Courts-Martial, and AR 381-10, Procedure 7, protect individuals against unreasonable searches and seizures of their persons, houses, papers, and effects and advises that this right will not be violated. The Fourth Amendment applies in our federal, state, and military court systems.

(1) An unlawful search is one made of a person, a person's house, papers, or effects without probable cause to believe that thereon or therein are located certain objects which are subject to lawful seizure. Probable cause to search exists when there is a reasonable belief that the person, property, or evidence sought is located in the place or on the person to be searched. It means more than "mere suspicion" or "good reason to suspect"(based, for example, on a preliminary or unsubstantiated report), but may be based on hearsay or other legally obtained information. The existence of probable cause to search permits an investigator to seek and obtain a search warrant or authorization from an appropriate

judicial or military authority and conduct the desired search. The existence of probable cause may also justify an immediate search without a warrant if exigent circumstances (for example, hot pursuit into a residence, or investigation centering around an operable, movable vehicle) give rise to a reasonable belief that a delay to obtain a warrant would result in removal or destruction of the evidence.

(2) Generally, evidence found as a result of an illegal search or seizure is inadmissible in a military trial and might well taint other evidence, thus precluding further judicial action if a timely motion is made to suppress. Seek advice from the local SJA on exceptions to this rule. The legality of each search necessarily depends on all of the facts in each situation. A search may be overt or covert. The following are types of legal searches:

(a) A search conducted in accordance with the authority granted by proper search warrant is lawful. The warrant must be issued from a court or magistrate having jurisdiction over the place searched. On the installation, this may be a military judge or magistrate. Military judges and magistrates may issue search authorizations based upon probable cause pursuant to Military Rule of Evidence 315(d) and AR 27-10.

(b) A search of an individual's person, the clothing worn, and of the property in immediate possession or custody is lawful when conducted as incident to the lawful apprehension of such person.

(c) A search is lawful when made under circumstances requiring immediate action to prevent the removal or disposal of property believed on reasonable grounds to be evidence of a crime. (d) A search is legal when made with the freely given consent of the owner in possession of the property searched. However, such consent must be the result of a knowing and willing waiver of the rights of the individual concerned and not the mere peaceful submission to apparent lawful authority. Circumstances may dictate the need to obtain written permission of the owner to avoid later denials that permission was freely given.

(d) A commanding officer having jurisdiction over property owned or controlled by the US and under the control of an armed force may lawfully authorize the search of such property. It is immaterial whether or not the property is located in the US or a foreign country. Such a search must also be based on probable cause.

(3) For most purposes, routine physical security inspections in accordance with AR 380-5, and routine investigations of military or civilian personnel when entering or leaving military areas are not considered to be searches but are treated as legitimate administrative inspections or inventories. Contraband may be seized any time.

(4) If possible, the CI agent should request, in writing, the authority to search and should state sufficient factual information to support a conclusion that an offense has occurred or will occur and evidence of the offense is located at the place sought to be searched. Permission to search should be granted by endorsement to such a request. The law of search and seizure must always be related to the actual circumstances; the advice of an SJA or legal officer should be obtained in any doubtful case. The following procedures, however, are valid for any search:

(a) The CI agent secures all available evidence that an offense has been committed and that property relating to the offense is located at a specific place.

(b) The CI agent submits this evidence to the person with authority to order a search of the place or property.

(c) If the place or property is located in a civilian community in the US, the evidence is submitted to the judge or court with authority to issue a search warrant. To obtain a search warrant from a civilian court, CI agents must establish liaison with local civilian police agencies that are authorized to request search warrants and perform the search. The supporting SJA frequently performs a military-civilian liaison function, and should be consulted when such warrants are desired.

(d) If the place or property is located in a foreign country or occupied territory and is owned, used, or occupied by persons subject to military law or to the law of war, the evidence is submitted to a commanding officer of the US Forces who has jurisdiction over personnel subject to military law or to the law of war.

(e) If the place where the property is owned or controlled by the US is under the control of an armed force wherever located, the evidence is submitted to the commanding officer having jurisdiction over the place where the property is located.

(f) The person with authority to order a search must find in the evidence probable cause to believe that the specified place or property contains specific objects subject to lawful seizure.

(g) If the person finds probable cause, that person may then lawfully authorize the search.

(h) Having been authorized, the CI agent may search the specified place or property for the specified objects.

(5) It is possible to have a legal seizure during an illegal search. (For example, the seizure of contraband is always legal, although the illegality of the search may prevent the use of such contraband as evidence.) It is possible to have an illegal seizure during a legal search. In any given judicial procedure, the first point of inquiry will be the legality of the search. If it was illegal, there will be no need to go any further; only if the search was legal will it become necessary to determine the legality of the seizure.

(6) If the search is lawful, certain objects may be seized and admitted in evidence against the suspect:

    (a) Contraband, such as property which is prohibited by law. Examples are drugs and untaxed liquor.

    (b) Fruits of the crime. Property which has been wrongfully taken or possessed.

    (c) Tools or means by which the crime was committed.

    (d) Evidence of the crime, such as clothing.

c. Apprehension. Apprehension (called "arrest" in many civilian jurisdictions) is the taking of a person into custody. A person has been taken into custody or apprehended when his or her freedom of movement is restricted in any substantial way.

    (1) Authorized individuals may apprehend persons subject to the UCMJ upon reasonable belief that an offense has been committed and that the person to be apprehended committed the offense. This is sufficient probable cause for an apprehension. The authority of the CI agent to apprehend is specified in the following documents: Article 7, UCMJ; Rule 302, Manual for Courts-Martial, 1984; and AR 381-20.

    (2) The basis for arrest by civilian police depends on the particular jurisdiction concerned. In general, civilian police make arrests either—

        (a) By a warrant upon a showing of probable cause to a magistrate.

        (b) Without a warrant, but for probable cause, when a felony or misdemeanor is committed or attempted in their presence.

        (c) If a reasonable belief exists that the person committed the offense.

    (3) Incident to a lawful apprehension, the SUBJECT's person, clothing that is worn, and property in the

immediate possession or control may be searched. Any weapon or means of escape may be lawfully seized.

d. Legal restrictions. For legal restrictions, see AR 381-10. For additional explanation and analysis see INSCOM Pamphlet 27-1. For explanations on proper handling of evidence, see AR 195-5 and FM 19-20. Questions should be referred to the serving SJA or legal advisor.

A-II-5. Intelligence Oversight.

a. AR 381-10 mandates that MI personnel conform to the spirit of the regulation in the conduct of intelligence collection. AR 381-10, Procedure 14, concisely states:

Employees shall conduct intelligence activities only pursuant to, and in accordance with, EO 12333 and this regulation. In conducting such activities, employees shall not exceed the authorities granted the employing DOD intelligence components by law; EO, including EO 12333, and the applicable DOD and Army directives.

b. Because of this requirement, intelligence personnel have the responsibility to understand the limits of the authority under which they conduct an activity, and the procedures from the regulation that apply to the given activity.

c. AR 381-10, Procedure 15, obligates each person to report any questionable intelligence activity by electrical message through command channels to HQDA (SAIG-IO). The term "questionable activity" refers to any conduct that constitutes or is related to an intelligence activity that may violate the law, any EO, or any DOD or Army policy including AR 381-10.

Section III
Technical Investigative Techniques
to
Appendix A
Counter-Human Intelligence Techniques
And Procedures

A-III-1. General.

The conduct of investigations is enhanced many times by emerging sophisticated procedures designed to simplify and shorten the time required to complete certain investigative tasks while ensuring that all evidence, no matter how seemingly insignificant, is thoroughly evaluated. CI units have available to them from higher supporting echelons and from within their own resources, personnel skilled in technical investigative techniques.

a. Technical investigative techniques can contribute materially to the overall investigation. Section III identifies how each technique contributes to the total CI effort. They can assist in supplying the commander with factual information on which to base decisions concerning the security of the command. Investigators can use these techniques in connection with CI and personnel security investigations for LAAs. CI investigators selectively use the following technical services:

   (1) Electronic surveillance.
   (2) Investigative photography.
   (3) Laboratory analysis.
   (4) Polygraph (authorized for use in local access investigations).
   (5) TEMPEST.
   (6) Computer CE capabilities.

b. In addition, specially trained CI agents conduct TSCM investigations to detect clandestine surveillance systems. TSCM personnel and intelligence polygraph examiners are also trained and experienced CI agents.

A-III-2. Electronic Surveillance.

Electronic surveillance is the use of electronic devices to monitor conversations, activities, sound, or electronic impulses. It is an aid in conducting investigative activities. The US Constitution; EO 12333; and AR 381-10, AR 381-14 (S), and AR 381-20 regulate the use of wiretapping and electronic eavesdropping.

a. Technical surveillance methodology, including those employed by FIS, consists of—

(1) Pickup devices. A typical system involves a trans-ducer (such as a microphone, video camera, or similar device) to pick up sound or video images and convert them to electrical impulses. Pickup devices are available in practically any size and form. They may appear to be common items, such as fountain pens, tie clasps, wristwatches, or household or office fixtures. It is important to note that the target area does not have to be physically entered to install a pickup device. The availability of a power supply is the major limitation of pickup devices. If the device can be installed so its electrical power is drawn from the available power within the target area, there will be minimal, if any, need for someone to service the device.

(2) Transmission links. Conductors carry the impulses created by the pickup device to the listening post. In lieu of conductors, the impulses can go to a transmitter which converts the electrical impulses into a modulated radio frequency (RF) signal for transmission to the listening post. The simplest transmission system is conventional wire. Existing conductors, such as used and unused telephone and electrical wire or unground electrical conduits, may also be used. The development of miniature electronic components permits the creation of very small, easily concealed RF transmitters. Such transmitters may operate from standard power sources or may be battery operated. The devices themselves may be continuously operated or remotely activated.

(3) Listening posts. A listening post consists of an area containing the necessary equipment to receive the signals from the transmission link and process them for monitoring or recording.

    (a) Listening posts use a receiver to detect the signal from an RF transmission link. The receiver converts the signal to an audio-video frequency and feeds it to the monitoring equipment. Use any radio receiver compatible with the transmitter. Receivers are small enough to be carried in pockets and may be battery operated.

    (b) For wire transmission links only, a tape recorder is required. You can use many commercially available recorders in technical surveillance systems. Some of these have such features as a voice actuated start-stop and variable tape speeds (extended play). They may also have automatic volume control and can be turned on or off from a remote location.

b. Monitoring telephone conversations is one of the most productive means of surreptitious collection of information. Because a telephone is used so frequently, people tend to forget that it poses a significant security threat. Almost all telephones are susceptible to "bugging" and "tapping."

    (1) A bug is a small hidden microphone or other device used to permit monitoring of a conversation. It also allows listening to conversations in the vicinity of the telephone, even when the telephone is not in use.

    (2) A telephone tap is usually a direct connection to the telephone line which permits both sides of a telephone conversation to be monitored. Tapping can be done at any point along the line, for example, at connector blocks, junction boxes, or the multiwire cables leading to a telephone exchange or dial central office. Telephone lineman's test sets and miniature telephone monitoring devices are examples of taps. Indirect tapping of a line, requiring no physical connection to the line, may also be accomplished.

(3) The most thorough check is not absolute insurance against telephone monitoring. A dial central office or telephone exchange services all telephone lines. The circuits contained within the dial central office allow for the undetected monitoring of telephone communications. Most telephone circuits servicing interstate communications depend on microwave links. Communications via microwave links are vulnerable to intercept and intelligence exploitation.

c. Current electronic technology produces technical surveillance devices that are extremely compact, highly sophisticated, and very effective. Miniaturized technical surveillance systems are available. They can be disguised, concealed, and used by a FIS in a covert or clandestine manner. The variations of their use are limited only by the ingenuity of the technician. Equipment used in technical surveillance systems varies in size, physical appearance, and capacity. Many are identical to, and interchangeable with, components of commercially available telephones, calculators, and other electronic equipment.

A-III-3. Investigative Photography and Video Recording.

a. Photography and video recording in CI investigations includes—

(1) Identification of individuals. CI agents perform both overt and surreptitious photography and video recording.

(2) Recording of incident scenes. Agents photograph overall views and specific shots of items at the incident scene.

(3) Recording activities of suspects. Agents use photography and video recording to provide a record of a suspect's activities observed during surveillance or cover operations.

b. A photograph or video recording may be valuable as evidence since it presents facts in pictorial form and creates realistic mental impressions. It may present

---

PHOTO DATA CARD

Case number: _____ Subject: _____ Photographer: _____
Location: _____ Date: _____
Time of day: _____ Weather conditions: _____
Camera: _____ Negative size: _____
Lens (type): _____ Focal length: _____
Shutter speed: _____ f- stop: _____
Film: _____
Camera position: _____
A. Compass reading: _____ B. Height and altitude: _____
C. Lateral position: _____ D.Tilt: _____
E. Camera-to-subject distance: _____
Artificial light used: _____ Developer: _____
Developing time: _____ Temperature: _____ Agitation: _____
Method of printing: _____ Contrast: _____
Type of enlarger lens: _____
Paper: _____
Distance between important objectives in view: _____
Description of area: _____
_____
_____
Remarks: _____
_____
_____

Figure A-III-1. Sample photo data card.

evidence more accurately than a verbal description. Photographs permit consideration of evidence which, because of size, bulk, weight, or condition, cannot be brought into the courtroom.

c. To qualify as evidence, photographs and video recordings must be relevant to the case and be free of distortion. A person who is personally acquainted with the locale, object, person, or thing represented must verify the photograph or video recording. This is usually the photographer. The agent will support photographs and video recordings used as evidence by notes made at the time of the photography. These notes provide a description of what the photograph includes. The notes will contain—

(1) The case number, name of the subject, and the time and date that the photographs or video recordings were taken.

110

(2) Technical data, such as lighting and weather conditions and type of film, lens, and camera used.

(3) Specific references to important obj ects in the photograph.

d. These notes may be retained on a form such as a photo data card shown in Figure A-III-1.

e. Agents can obtain specialized photographic development support from the Intelligence Materiel Activity, Fort Meade, MD. Physical surveillance of US persons including photography and video recording, is governed by AR 381-10, Procedure 9.

A-III-4. Laboratory analysis.

We must anticipate the use of false documentation and secret writing by foreign intelligence agents in many CI investigations. Detection requires specially trained personnel and laboratory facilities. The CI unit SOP should list how this support is obtained.

A-III-5. Polygraph.

a. The polygraph examination is a highly structured technique conducted by specially trained CI technicians and civilians certified by proper authority as polygraph examiners. Provisions of AR 195-6 cover the polygraph program generally; AR 381-20 covers intelligence polygraphs.

(1) AR 195-6 describes general applicability, responsibilities, and use of polygraph; records processing; and selection and training of DA polygraph examiners.

(2) AR 381-20 authorizes intelligence polygraphs for CI investigations, foreign intelligence and CI operations, personnel security investigations, access to SCI, exculpation in CI and personnel security investigations; and CI scope polygraph (CSP) examinations in support of certain programs or activities listed in AR 381-20.

b. The conduct of the polygraph examination is appropriate, with respect to investigations, only when—

(1) All investigative leads and techniques have been completed as thoroughly as circumstances permit.

(2) The subject of the investigation has been interviewed or thoroughly debriefed.

(3) Verification of the information by means of polygraph is deemed essential for completion or continuation of the investigation.

c. Do not conduct a polygraph examination as a substitute for securing evidence through skillful investigation and interrogation. The polygraph examination is an investigative aid and can be used to determine questions of fact, past or present. CI agents cannot make a determination concerning an individual's intentions or motivations, since these are states of mind, not fact. However, consider the examination results along with all other pertinent information available. Polygraph results will not be the sole basis of any final adjudication.

d. Conduct polygraph examinations during CI and personnel security investigations to—

(1) Determine if a person is attempting deception concerning issues involved in an investigation.

(2) Obtain additional leads concerning the facts of an offense, the location of items, whereabouts of persons, or involvement of other, previously unknown individuals.

(3) Compare conflicting statements.

(4) Verify statements from witnesses or subjects to include CI and personnel security investigations.

(5) Provide a just and equitable resolution of a CI or personnel security investigation when the subject of such an investigation requests an exculpatory polygraph in writing.

e. Conduct intelligence polygraph examinations to—

(1) Determine the suitability, reliability, or creditability of agents, sources, or operatives of foreign intelligence or CI operations.

(2) Determine the initial and continued eligibility of individuals for access to programs and activities authorized CSP examination support.

f. The polygraph examination consists of three basic phases: pretest, intest, and posttest.

(1) During the pretest, appropriate rights advisement are given and a written consent to undergo polygraph examination is obtained from all examinees who are suspects or accused. Advise the examinee of the Privacy Act of 1974 and the voluntary nature of examination. Conduct a detailed discussion of the issues for testing and complete the final formulation of questions to be used during testing.

(2) During the intest phase, ask previously formulated and reviewed test questions and monitor and record the examinee's responses by the polygraph instrument. Relevant questions asked during any polygraph examination must deal only with factual situations and be as simple and direct as possible. Formulate these questions so that the examinee can answer only with a yes or no. Never use or ask unreviewed questions during the test.

(3) If responses indicate deception, or unclear responses are noted during the test, conduct a posttest discussion with the examinee in an attempt to elicit information from the examinee to explain such responses.

g. A polygraph examiner may render one or more of four possible opinions concerning the polygraph examination.

(1) No opinion (NO) is rendered when less than two charts are conducted concerning the relevant issues, or a medical reason halts the examination. Normally, three charts are conducted.

(2) Inconclusive (INCL) is rendered when there is insufficient information upon which to make a determination.

(3) No deception indicated (NDI) is rendered when responses are consistent with an examinee being truthful regarding the relevant areas.

(4) Deception indicated (DI) when responses are consistent with an examinee being untruthful to the relevant test questions.

h. Certain mental or physical conditions may influence a person's suitability for polygraph examination and affect responses during testing. CI agents should report any information they possess concerning a person's mental or physical condition to the polygraph examiner before scheduling the examination. Typical conditions of concern are—

(1) Mental disorders of any type.
(2) Any history of heart, respiratory, circulatory, or nervous disorders.
(3) Any current medical disorder, to include colds, allergies, or other conditions (such as pregnancy or recent surgery).
(4) Use of drugs or alcohol before the examination.
(5) Mental or physical fatigue.
(6) Pain or physical discomfort.

i. To avoid such conditions as mental or physical fatigue, do not conduct prolonged or intensive interrogation or questioning immediately before a polygraph examination. The CI agent tells the potential examinee to continue taking any prescribed medication and bring it to the examination. Based on information provided by the CI agent and the examiner's own observations, the polygraph examiner decides whether or not a person is fit to undergo examination by polygraph. When the CI agent asks a person to undergo a polygraph examination, the person is told that the examination is voluntary and that no adverse action can be taken based solely on the refusal to undergo examination by polygraph. Further, the person is informed that no information concerning a refusal to take a polygraph examination is recorded in any personnel file or record.

j. The CI agent will make no attempt to explain anything concerning the polygraph instrument or the conduct of the examination. If asked, the CI agent should inform the person that the polygraph examiner will provide a full explanation of the instrument and all procedures before actual testing and that all test questions will be fully reviewed with the potential examinee before testing.

k. Conduct polygraph examinations in a quiet, private location. The room used for the examination must contain, as a minimum, a desk or table, a chair for the examiner, and a comfortable chair with wide arms for the examinee. The room may contain minimal, simple decorations; must have at least one blank wall; and must be located in a quiet, noise-free area. Ideally, the room should be soundproof. Visual or audio monitoring devices may be used during the examination; however, the examiner must inform the examinee that such equipment is being used and whether the examination will be monitored or recorded in any manner.

l. Normally, only the examiner and the examinee are in the room during examination. When the examinee is an accused or suspect female and the examiner is a male, a female witness must be present to monitor the examination. The monitor may be in the examination room or may observe through audio or visual equipment, if such is available.

m. On occasion, the CI agent must arrange for an interpreter to work with the examiner. The interpreter must be fluent in English and the required language, and have a security clearance appropriate to the classification of material or information to be discussed during the examination. The interpreter should be available in sufficient time before the examination to be briefed on the polygraph procedures and to establish the proper working relationship.

n. AR 195-6 describes polygraph reports, records to be maintained, and records distribution. The CI agent must provide the examiner with all files, dossiers, and reports pertaining to the investigation or operation before the examination, and must be available to answer any questions the examiner may have concerning the case.

(1) The CI agent will not prepare any agent reports concerning the results of a polygraph examination. This does not include information derived as a result of pretest or posttest admissions, nor does it include those situations where the CI agent must be called upon by the examiner to question the subject concerning those areas which must be addressed before the completion of the examination.

(2) The polygraph examiner will prepare a DA Form 2802. A copy of this may be provided to the CI agent. Such copies must be destroyed within three months following completion of the investigation or operation. The Investigative Records Repository, Central Security Facility, Fort Meade, MD, maintains the original of the DA Form 2802. Request polygraph support in accordance with INSCOM Pamphlet 381-6.

A-III-6. Technical Surveillance Countermeasures.

a. TSCM versus TEMPEST. TSCM is concerned with all signals leaving a sensitive or secure area, to include audio, video, and digital or computer signals. There is a definite distinction between TSCM and TEMPEST.

(1) TEMPEST is the unintentional emanation of electronic signals outside a particular piece of equipment. Electric typewriters create such signals. The words to focus on in TEMPEST are "known" and "unintentional" emanations. TEMPEST is controlled by careful engineering or shielding.

(2) TSCM is concerned with the intentional effort to gather intelligence by foreign intelligence activities by impulsing covert or clandestine devices into a US facility, or modifying existing equipment within that area. For the most part, intelligence gained through the use of technical surveillance means will be accurate, as people are unaware they are being monitored. At the same time, the

implanting of such technical surveillance devices is usually a last resort.

b. Threat. The FIS, their agents, and other persons use all available means to collect sensitive information. One way they do this is by using technical surveillance devices, commonly referred to as "bugs" and "taps." Such devices have been found in US facilities worldwide. Security weaknesses in electronic equipment used in everyday work have also been found worldwide. The FIS easily exploits these weaknesses to collect sensitive or classified conversations as well as the information being processed. They are interested in those things said in (supposed) confidence, since they are likely to reveal future intentions. It should be stressed that the threat is not just audio, but video camera signals, as well as data. Devices are usually placed to make their detection almost impossible without specialized equipment and trained individuals.

c. The TSCM program. The purpose of the TSCM program is to locate and neutralize technical surveillance devices that have been targeted against US Government sensitive or secure areas. The TSCM program is designed to identify and enable the correction of exploitable technical and physical security vulnerabilities. The secondary, and closely interrelated purpose, is to provide commanders and department heads with a comprehensive evaluation of their facilities' technical and physical security postures. The Director of Central Intelligence established the requirement for a comprehensive TSCM program. DODD 5240.5 and AR 381-14 (S) govern the implementation of this program.

(1) The TSCM program includes four separate functions; each with a direct bearing on the program.

(a) Detection. Realizing that the threat is there, the first and foremost function of the TSCM program is to detect these devices. Many times these devices cannot be easily detected. Occasionally, TSCM personnel will discover such a device by accident. When they discover a device, they must neutralize it.

(b) Nullification. Nullification includes both passive and active measures used to neutralize or negate devices that are found. An example of passive nullification is soundproofing. But soundproofing that covers only part of a room is not very helpful. Excessive wires must be removed, as they could be used as a transmission path from the room. Nullification also refers to those steps taken to make the emplacement of technical surveillance systems as difficult as possible. An example of active nullification is the removal of a device from the area.

(c) Isolation. The third function of the TSCM program is isolation. This refers to limiting the number of sensitive or secure areas and ensuring the proper construction of these areas.

(d) Education. Individuals must be aware of the foreign intelligence threat and what part they play should a technical surveillance device be detected. Additionally, people need to be alert to what is going on in and around their area, particularly during construction, renovations, and installation of new equipment.

(2) The TSCM program consists of CI technical investigations and services (such as surveys, inspections, preconstruction advice and assistance) and technical security threat briefings. TSCM investigations and services are highly specialized CI investigations and are not to be confused with compliance-oriented or administrative services conducted to determine a facility's implementation of various security directives.

(a) TSCM survey. This is an all-encompassing investigation. This investigation is a complete electronic, physical, and visual examination to detect clandestine surveillance systems. A byproduct of this investigation is the identification of physical and technical security weaknesses which could be exploited by the FIS.

(b) TSCM inspection. Normally, once a TSCM survey has been conducted, it will not be repeated. If TSCM personnel note several technical and physical weaknesses during the survey, they may request and schedule an inspection at a later date. In addition, they will schedule an inspection if there has been an increased threat posed to the facility or if there is some indication that a technical penetration has occurred in the area. DODD 5240.5 specifically states that no facility will qualify automatically for recurrent TSCM support.

(c) TSCM preconstruction assistance. As with other technical areas, it is much less expensive and more effective to build in good security from the initial stages of a new project. Thus, preconstruction assistance is designed to help security and construction personnel with the specific requirements needed to ensure that a building or room will be secure and built to standards. This saves money by precluding costly changes later on.

(3) Army activities request TSCM support in accordance with AR 381-14 (S).

(a) Requests for, or references to, a TSCM investigation will be classified SECRET and marked with the protective marking, NOT RELEASABLE TO FOREIGN NATIONALS. The fact that support is scheduled, in progress, or completed, is classified SECRET.

(b) No request for TSCM support will be accepted via nonsecure means. Nonsecure telephonic discussion of TSCM support is prohibited.

(c) All requests will be considered on a case-by-case basis and should be forwarded through the appropriate major Army command to

Commanding General, US Army Intelligence and Security Command, ATTN: IAOPS-CI-TC, Fort Belvoir, VA 22060-5246.

(d) When requesting or receiving support, the facility being inspected must be complete and operational, unless requesting preconstruction advice and assistance. If any additional equipment goes into the secure area after the investigation, the entire area is suspect and the investigation negated.

(e) Fully justified requests of an emergency nature, or for new facilities, may be submitted at any time, but should be submitted at least 30 days before the date the support is required. Unprogrammed requests will be funded by the requestor. Each request for unprogrammed TSCM support must be accompanied by a fund cite to defray the costs of temporary duty (TDY) and per diem.

(4) The compromise of a TSCM investigation or service is a serious security violation with potentially severe impact on national security. Do not compromise the investigation or service by any action which discloses to unauthorized persons that TSCM activity will be, is being, or has been conducted within a specific area. Unnecessary discussion of a TSCM investigation or service, particularly within the subject area, is especially dangerous.

(a) If a listening device is installed in the area, such discussion can alert persons who are conducting the surveillance and permit them to remove or deactivate their devices. When deactivated, such devices are extremely difficult to locate and may require implementation of destructive search techniques.

(b) In the event a TSCM investigation or service is compromised, the TSCM team chief will terminate the investigation or service at once. Report the circumstances surrounding the compromise of the investigation or service to the head of the serviced facility, the appropriate major Army command, and the INSCOM TSCM Program Director. TSCM personnel will not reschedule an investigation or service until the cause and impact of the compromise have been evaluated by the TSCM CI agent, the appropriate agency head, and the INSCOM TSCM Program Director.

(5) When a TSCM survey or inspection is completed, the requestor is usually given reasonable assurance that the surveyed area is free of active technical surveillance devices or hazards.

(a) TSCM personnel inform the requestor about all technical and physical security vulnerabilities with recommended regulatory corrective actions.

(b) The requestor should know that it is impossible to give positive assurance that there are no devices in the surveyed area.

(c) The security afforded by the TSCM investigation will be nullified by the admission to the secured area of unescorted persons who lack the proper security clearance. The TSCM investigation will also be negated by—

    i.   Failing to maintain continuous and effective surveillance and control of the serviced area.

    ii. Allowing repairs or alterations by persons lacking the proper security clearance or not under the supervision of qualified personnel.

    iii. Introducing new furnishings or equipment without a thorough inspection by qualified personnel.

(6) Report immediately the discovery of an actual or suspected technical surveillance device via a secure means, in accordance with guidance provided in AR 381-14 (S). All information concerning the discovery will be handled at a minimum of SECRET. Installation or unit security managers will request an immediate investigation by the supporting CI unit or supporting TSCM element.

Section IV
Screening, Cordon, and Search Operations
to
Appendix A
Counter-Human Intelligence Techniques
And Procedures

A-IV-1. General.

Screening, cordon, and search operations are used to gain intelligence information. Section IV provides the techniques and procedures for these operations. Screening operations identify individuals for further interrogation by CI and inter-rogators. CI agents may conduct screening at MP roadblocks or checkpoints. Cordon and search operations identify and apprehend persons hostile to our operations. Actual control-ling of areas is done by host nation forces assisted by CI, interrogation, and other friendly forces. In some instances, the operation may be exclusively US.

a. In a conventional combat environment, CI screening operations screen refugees, EPW, and civilian internees at mobile and static checkpoints. CI agents normally conduct these operations with other elements such as MP, interrogators, combat troops, CA, and PSYOP teams. These operations require close coordination and planning. The planning may include joint or combined planning. CI exploits cordon and search operations for individuals and information of CI interest, but is not in charge. The commander of the unit performing the cor-don and search is in charge.

b. In OOTW, CI agents use cordon and search operations to ferret out the insurgent infrastructure as well as individual unit elements which may use a community or area as cover for their activities or as a support base. CI agents conduct these operations, whenever possible, with host country forces and organizations.

c. Ideally, US Forces, including CI personnel, provide support while host country officials direct the entire operation. Host country personnel should, as a minimum, be part of the screening and sweep elements on any cordon and search operation. In situations where there is no viable host nation government, these operations may be conducted unilaterally or as part of a combined force.

A-IV-2. Counterintelligence Screening.

The purpose of CI screening operations is to identify persons of CI interest or verify persons referred by interrogators who are of CI interest, and gather information of immediate CI interest.

a. Subjects of intelligence interest. Interrogators normally conduct refugee and EPW screening at the EPW compound or refugee screening point. Interrogators refer persons identified for possible CI interest to CI personnel to be screened. CI personnel conduct interrogations with the view to intercepting hostile intelligence agents, saboteurs, and subversives trying to infiltrate friendly lines. As the battle lines in combat change, entire segments of the population may be overrun. The local population in any area may be swelled by refugees and displaced persons (persons from other lands conscripted by enemy forces for labor). The following are examples of categories of persons of CI interest (this list is not all inclusive):

(1) Persons suspected of attempting to infiltrate through refugee flow.
(2) Line crossers.
(3) Deserters from enemy units.
(4) Persons without identification (ID) papers or forged papers (inconsistent with the norm).
(5) Repatriated prisoners of war and escapees.
(6) Members of underground resistance organizations seeking to join friendly forces.
(7) Collaborators with the enemy.
(8) Target personalities, such as those on the personalities list (also known as the black, gray, or white lists).

(9) Volunteer informants.

(10) Persons who must be questioned because they are under consideration for employment with US Forces or for appointment as civil officials by CA units.

b. Planning and coordination. CI personnel plan these screening operations, as far as possible, in conjunction with the following elements:

(1) Combat commander. The commander is concerned with channelizing refugees and EPWs through the AO, particularly in the attack, to prevent any hindrance to unit movement, or any adverse effect on unit mission.

(2) Interrogators. Interrogation personnel must understand what CI is looking for and have the commander's current PIR and information require- ments (IR). Close coordination with interrogators is essential for successful CI operations.

(3) Military police. MP elements are responsible for collecting EPW and civilian internees from captur- ing units as far forward as possible in the AO. MP units guard the convoys transporting EPW and civilian internees to EPW camps, and command and operate the EPW camps.

(4) Civil affairs. CA elements, under the G5, are responsible for the proper disposition of refugees.

(5) Psychological operations. PSYOP elements, under the G3, contribute to screening operations by informing the populace of the need for their displacement.

(6) Civil authorities in hostile areas. Civil authorities in hostile areas are included in planning only if con- trol has been returned to them.

c. Preparation. Before any screening operation, the CI teams involved should become intimately familiar with all available information or indicators as covered in paragraph A-IV-2f, as well as the following facts:

(1) Regulations. To have any success, CI personnel must become familiar with all restrictions placed on the civilian population within the enemy-held area,

including curfews, travel restrictions, rationing, draft or conscription regulations, civilian labor force mobilization orders, required political organizational membership. Knowledge of these regulations may help the CI screener to detect discrepancies and discern changes in enemy activity.

(2) Intelligence, infrastructure, organization. In order to identify agents of the enemy intelligence or infrastructure apparatus, CI personnel must be thoroughly familiar with their methods of operation, policies, objectives, offices and suboffices, schools, officials, and known agents. This includes knowing what the enemy calls itself and its organization as well as the known names for the same organization.

(3) Order of battle. The CI team needs to maintain and have ready access to current OB information. All team members must know what adversary forces they are facing and what units are in the AO. CI teams need to know adversary unit disposition, composition, strength, weaknesses, equipment, their training, history, activities, and personalities. The better CI teams know their adversaries, the better they can employ the principles of CI in support of their own unit's mission.

(4) Area of operations. CI personnel should also become familiar with the area in which they are operating; particularly geography, landmarks, distances, and travel conditions. Knowing such pertinent information as the political situation, social and economic conditions, customs, and racial problems of the area is essential.

(5) Lists and information sheets. CI teams should distribute apprehensions lists and information sheets listing indicators of CI interest to the troops, MP, or other personnel assisting with the screening operation. CI teams should make up forms and pass them out to the individuals to be screened requiring them to record personal data. This form will aid in formulating the type of questions to be asked and in determining the informational areas needed to fulfill PIR and IR. Include the following data, plus anything else judged necessary, on the form:

    (a) Full name, other names, date and place of birth, current and permanent residences, and current citizenship.

    (b) The same information as above concerning the father, mother, and siblings, including the occupation and whereabouts of each.

    (c) If married, the names of spouse (including female maiden name), date, place of birth (DPOB), nationality, occupation, and personal data on spouse's family.

    (d) The individual's education and knowledge of languages.

    (e) Details of the individual's career to include schools, military service, technical and professional qualifications, political affiliations, and countries visited.

    (f) Point of departure, destination, and purpose.

NOTE: The Geneva Conventions do not require this, and if the person refuses to give the information, there is nothing that can be done about it. Prepare the form in the native language of the host nation and enemy force, if different. Ensure that it is prepared in the proper dialect of the language.

d. Main battle area screening.

    (1) Capturing troops search EPWs and internees captured in the main battle area (MBA) for weapons and documents, and prepare EPW and civilian internee capture tags. MP tasked with EPW operations collect EPW and civilian internees from capturing units as far forward as possible, normally establishing a division forward EPW and civilian internee collection point in or near the brigade support area (BSA). MP are responsible for subsequent evacuation of EPWs and civilian internees to the division central collection point and further rearward to internment sites.

    (2) Initial screening should take place at the brigade collecting points. The initial screening will, as a

minimum, consist of interrogation by intelligence interrogation personnel. Interrogation does not take precedence over rapid evacuation of EPW and civilian internees from dangerous areas. This is required by Article 19, Geneva Convention. Segregate individuals of CI interest from other EPW and civilian internees. Identify them to evacuating MP who will refer them to a CI team at division. EPW and civilian internees of CI interest remain segregated and are referred to CI teams for coordination of more detailed interrogation as they pass through the evacuation process.

e. Conduct. Because of time and the large numbers of people to be interrogated, it is impossible to interrogate everyone. Civilians moving about the combat area have to be subjected to brief inquiries on a selective basis by MI, CA, PSYOP, and MP personnel. Such brief inquiries are designed to locate and separate suspicious persons from the masses and should be thought of as a preliminary interrogation.

(1) While some are detained for interrogation, some selected persons are detained for further CI interrogation. Upon notification of a detainee or prisoner of CI interest, a CI team will be dispatched as soon as possible to the collection or screening point. The CI team will then coordinate with the interrogation team to determine the best method for conducting the CI interrogation or screening. If a determination is made that the EPW or detainee is of CI interest, the CI team will either control operational activity or refer the operation to the next higher echelon. If the detainee is to be referred to a higher echelon for the detailed interrogation, furnish a preliminary screening sheet and SPOT report to the evacuating unit. The evacuating unit will deliver the detainee and screening report to the next echelon CI team.

(2) The CI screening report should include the following:

128

(a) Identity . Screen all identifying documents in the form of ID cards, ration cards, draft cards, driver's license, auto registration, travel documents, and passport. Record rank, service number, and unit if a person is, or has been a soldier. Check all this information against the form previously filled out by the detainee if this was done.

(b) Background. The use of the form identified earlier will aid in obtaining the information required; however, certain information areas on the forms will have to be clarified, especially if data indicate a suspect category or the person's knowledgeability of intelligence information. If the form has not been filled out at this point, try to gain the information through questioning.

(c) Recent activities. Examine the activities of persons during the days before their detainment or capture. What were they doing to make a living? What connection, if any, have they had with the enemy? Why were they in the area? This line of questioning may bring out particular skills such as those associated with a radio operator, linguist, or photographer. Make physical checks for certain types of calluses, bruises, or stains to corroborate or disprove his story. Sometimes soil on shoes will not match that from the area he claims to come from.

(d) Journey or escape route. CI personnel should determine the route the individual took to get to US lines or checkpoints. Question the individual further on time, distance, and method of travel to determine whether or not the trip was possible during the time stated and with the mode of transportation used. Discrepancies in travel time and distances can be the key to the discovery of an infiltrator with a shallow cover story. By determining what an individual observed enroute, the screener can either check the person's story or pick up intelligence information concerning the enemy forces. Interrogators are well trained in this process and should be called upon for assistance and training.

f. Indicators.

  (1) Use the following indicators in an attempt to identify hostile infiltrators. CI personnel look for persons:

    (a) Of military age.

    (b) Traveling alone or in pairs.

    (c) Without ID.

    (d) With unusual documents.

    (e) Possessing large amounts of money, precious metals, or gems.

    (f) Displaying any peculiar activity.

    (g) Trying to avoid detection or questioning.

    (h) Using enemy methods of operating.

    (i) Having a pro-enemy background.

    (j) With a suspicious story.

    (k) With a family in enemy areas.

    (l) With a technical skill or knowledge.

    (m) Who have collaborated.

    (n) Who violate regulations in enemy areas.

  (2) In addition to interrogation, use the following methods of screening EPWs and refugees:

    (a) Apprehension lists.

    (b) Low-level informants infiltrated into EPW compounds or camps; civilian internee screens or camps; or refugee screens or centers.

    (c) Sound equipment placed in suspect-holding areas or cages.

    (d) Polygraph examinations.

(e) Specialized identification equipment, for example, metal-trace detection kits.

g. Checkpoints.

(1) This type of CI screening requires CI personnel to prepare apprehension lists and indicators to be used by screening teams. Specialized equipment such as metal detection kits would significantly enhance the screening process. These teams will provide the initial screening and will detain and refer suspects to the MI control element for detailed OB or CI interrogation and possible exploitation. Screening teams can be made up of combat troops, MP, CA, intelligence interrogators, and CI agents.

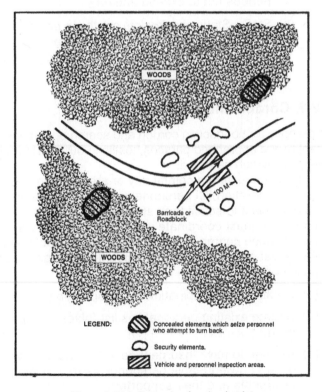

Figure A-IV-1. Example of a checkpoint.

(2) Place checkpoints shown in Figure A-IV-1 at strategic locations, where there is sufficient space for assembling people under guard and for parking vehicles for search and investigation. Set these up as either mobile or static missions. Post local security to protect the checkpoint and post a sufficient amount of personnel to the front and rear to catch anyone attempting to avoid the checkpoint. The preparation needed for static and mobile checkpoints is identical to other screening operations, and the indicators will remain basically the same.

    (a) Mobile. Use a mobile checkpoint as a moving system by which the team, either mounted or on foot, briefly selects individuals at random. Locate these checkpoints at various points for periods not to exceed one day.

    (b) Static. Static checkpoints are those manned permanently by MP or troops at the entrance to a bridge, town gate, river crossing, or similar strategic point.

A-IV-3. Cordon and Search.

The purpose for conducting cordon and search operations is to identify and apprehend persons hostile to our efforts and to exploit information gathered.

a. Before conducting a community or area cordon and search operation, CI personnel must coordinate with local officials to solicit their support and cooperation. They must coordinate with the host country area coordination center, if established. If not established, they must coordinate with host country intelligence and police organization to—

    (1) Obtain their participation in the operation.

    (2) Update existing personalities list (black and gray lists).

    (3) Arrange to have insurgent defectors, agents, and other knowledgeable personnel present to identify insurgents and their supporters.

(4) Update all intelligence on the community or area.

b. CI personnel must coordinate with appropriate US and host country CA and PSYOP units. Coordination must also be done with the unit conducting the operation. An essential part of preparing for a cordon and search is an update of all intelligence on the community or area.

c. The senior tactical unit commander will be the individual responsible for the conduct of the operation. That commander will plan, with advice from CI, interrogation, CA, and PSYOP personnel, the cordon which is usually deployed at night, and the search which normally begins at first light.

d. Community operations.

(1) The basic operation is the community cordon and search operation shown in Figure A-IV-2. As the screening element sets up the collection or screening station shown in Figure A-IV-3, the sweep element escorts the residents toward the station,

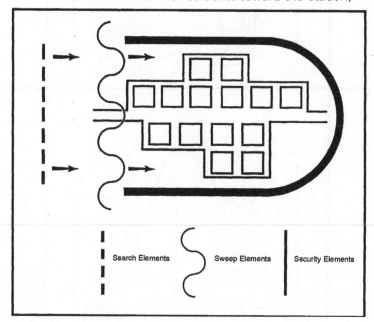

Search Elements   Sweep Elements   Security Elements

Figure A-IV-2. Example of community cordon and search operations.

133

leaving behind one resident to care for family belongings, if required by law.

(2) The search element follows behind the sweep element searching houses, storage areas, cemeteries and so forth, with dogs and metal detection equipment. CI personnel are searching for evidence of intelligence collection operations to include communications codes or other such paraphernalia. Each search element should include a CI team with an interrogator team as required, which will have a list of persons of CI interest.

(3) In the collection or screening station, bring the residents to the collection area (or holding area) and then systematically lead them to specific screening stations. Enroute to the screening station, search each individual for weapons. Then lead

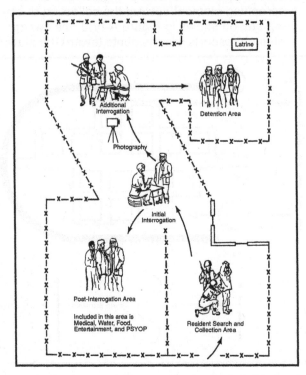

Figure A-IV-3. Example of collecting screening station.

the residents past the mayor or community leaders (enemy defectors or cooperating prisoners who will be hidden from view so that they can uncompromisingly identify any recognizable enemy). These informants will be provided with the means to notify a nearby guard or a screener if they spot an enemy member. Immediately segregate this individual and interrogate by appropriate personnel.

(4) At specific screening stations, ask the residents for identification, check against personalities list (black list), and search for incriminating evidence by electronic equipment.

(5) Move suspected persons on for photographing, further interrogation, or put them in the screening area detention point to be taken back to a base area or area coordination center interrogation facility for detailed interrogation upon completion of the operation.

(6) Pass innocent residents through to the post screening area where they are provided medical assistance and other civic assistance, as well as entertainment and friendly propaganda.

(7) Return any persons caught attempting to escape or break through the cordon immediately to the detention area.

(8) When the operation is terminated, allow all innocent individuals to return to their homes, and remove the enemy suspects under guard for further interrogation. Photograph all members of the community for compilation of a village packet, which will be used in future operations.

e. "Soft" or area operation.

(1) The second type of cordon and search operation is very frequently referred to as the "soft" or area cordon and search. This operation includes the cordoning and searching of a rather vast area (for example, a village area incorporating a number of hamlets, boroughs, towns, or villages which are

subdivisions of a political area beneath country level).

(2) This type of operation requires a multibattalion military force to cordon off the area; a pooling of all paramilitary, police, CA, and intelligence resources to conduct search and screening; and a formidable logistical backup. This kind of operation extends over a period of days and may take as long as a week or possibly longer.

(3) While screening and search teams systematically go from community to community and screen all residents, military forces sweep the area outside the communities over and over again to seek out anyone avoiding screening. As each resident is screened, CI agents will issue documents testifying to the fact that he was screened and if necessary, allow him restricted travel within the area.

(4) Other population and resources control measures are used as well. Such an opportunity may allow the chance to issue new ID cards and photograph all of the area's residents.

(5) As each community screening proceeds, send individuals who were designated for further interrogation to a centralized interrogation center in the cordoned area. Here, CI personnel will work with intelligence interrogation personnel, both US and indigenous, police, and other security service interrogators.

(6) Besides field files and other expedient facilities, a quick reaction force is located at the interrogation center to react immediately to intelligence developed during the interrogations and from informants planted among the detainees.

Section V
Personalities, Organizations, and Installations List
to
Appendix A
Counter-Human Intelligence Techniques and Procedures

A-V-1. General.

The effectiveness of CI operations depends largely on the planning that precedes the operation. Early in the planning process the CI officer on the G2 staff directs the efforts to obtain information on the adversary's intelligence, sabotage, terrorism, and subversion capabilities. Collected information is processed and analyzed, and from it the CI officer formulates a list of CI targets. Section V identifies the criteria for the personalities, organizations, and installations list. CI targets are personalities, organizations, and installations of intelligence or CI interest which must be seized, exploited, or protected.

A-V-2. Personalities.

These are persons who are a threat to security, whose intentions are unknown, or who can assist the intelligence and CI efforts of the command. Personalities are grouped into these three categories. For ease in identification, a color code indicates the category. Colors currently in use are black, gray, and white and pertain to the three categories in the order listed above.

a. Black list. The black list is an official CI listing of actual or potential enemy collaborators, sympathizers, intelligence suspects, and other persons whose presence menaces the security of the friendly forces. (Joint Pub 1-02) Black list includes—

    (1) Known or suspected enemy or hostile espionage agents, saboteurs, terrorists, political figures, and subversive individuals.

(2) Known or suspected leaders and members of hostile paramilitary, partisan, or guerrilla groups.

(3) Political leaders known or suspected to be hostile to the military and political objectives of the US or an allied nation.

(4) Known or suspected officials of enemy governments whose presence in the theater of operations poses a security threat to the US Forces.

(5) Known or suspected enemy collaborators and sympathizers whose presence in the theater of operations poses a security threat to the US Forces.

(6) Known enemy military or civilian personnel who have engaged in intelligence, CI, security, police, or political indoctrination activities among troops or civilians.

(7) Other personalities indicated by the G2 as automatic arrestees. Included in this category may be local political personalities, police chiefs, and heads of significant municipal and national departments or agencies, and tribal or clan leaders.

b. Gray list. The gray list contains the identities and locations of those personalities whose inclinations and attitudes toward the political and military objective to the US are obscure. Regardless of their political inclinations or attitudes, personalities may be listed on gray lists when they are known to possess information or particular skills required by US Forces. These people are the "unknowns." They may be individuals whose political motivations require further exploration before they can be used effectively by US Forces. Examples of individuals who may be included in this category are—

(1) Potential or actual defectors from the hostile cause whose bona fides have not been established.

(2) Individuals who have resisted, or are believed to have resisted, the enemy government and who may be willing to cooperate with US Forces, but whose bona fides have not been established.

(3) Scientists and technicians suspected of having been engaged against their will in enemy research projects of high technology programs.

c. White list. The white list contains the identities and locations of individuals who have been identified as being of intelligence or CI interest and are expected to be able to provide information or assistance in existing or new intelligence AIs. They are usually in accordance with, or favorably inclined toward, US policies. Their contributions are based on a voluntary and cooperative attitude. Decisions to place individuals on the white list may be affected by the combat situation, critical need for specialists in scientific fields, and such theater intelligence needs as may be indicated from time to time. Examples of individuals who may be included in this category are—

(1) Former political leaders of a hostile state who were deposed by the hostile political leaders.

(2) Intelligence agents employed by US or allied intelligence agencies.

(3) Key civilians in areas of scientific research, who may include faculty members of universities and staffs of industrial or national research facilities, whose bona fides have been established.

(4) Leaders of religious groups and other humanitarian groups.

(5) Other persons who can materially and significantly aid the political, scientific, and military objectives of the US and whose bona fides have been established.

A-V-3. Installations.

Installations on the CI targets list are any building, office, or field position that may contain information or material of CI interest or which may pose a threat to the security of the command. Installations of CI interest include—

a. Those that are or were occupied by enemy espionage, sabotage, or subversive agencies or police organizations, including prisons and detention centers.

b. Those occupied by enemy intelligence, CI, security, or paramilitary organizations including operational bases, schools, and training sites.

c. Enemy communication media and signal centers.

d. Nuclear research centers and chemical laboratories.

e. Enemy political administrative HQ.

f. Public utilities and other installations to be taken under early control to prevent sabotage.

g. Production facilities, supply areas, and other installations to be taken under control to prevent support to hostile guerrilla and partisan elements.

h. Embassies and consulates of hostile governments.

A-V-4. Organizations.

Any group that is a potential threat to the security of the friendly force must be neutralized, rendered ineffective. Groups or organizations which are of concern to CI during tactical operations include—

a. Hostile intelligence, sabotage, subversive, and insurgent organizations.

b. National and local political parties or groups known to have aims, beliefs, or ideologies contrary or in opposition to those of the US.

c. Paramilitary organizations, including students, police, military veterans, and excombatant groups known to be hostile to the US.

d. Hostile sponsored organizations or groups whose objectives are to create dissention and spread unrest among the civilian population in the AO.

A-V-5. Control Measures.

The CI officer and the G2 need a positive way to keep track of the status of CI targets. Called a target reduction plan, it's a checklist used to ensure targets are seized, exploited, or controlled in a timely manner. The plan is keyed to the scheme of maneuver and lists targets as they are expected to appear. When more targets appear than can be exploited, a priority list is used to denote which target takes priority.

a. Priority one targets represent the greatest threat to the command. They possess the greatest potential source of information or material of intelligence or CI value. Priority one targets must be exploited or neutralized first.

b. Priority two targets are of lesser significance than priority one. They are taken under control after priority one targets have been exploited or neutralized.

c. Priority three targets are of lesser significance than priority one or two. They are to be exploited or neutralized as time and personnel permit.

Section VI
Counter-Human Intelligence Analysis
to
Appendix A
Counter-Human Intelligence Techniques And Procedures

A-VI-1. General.

C-HUMINT analysis increases in importance with each new US involvement in worldwide operations. Especially in OOTW, C-HUMINT analysis is rapidly becoming a cornerstone upon which commanders base their concepts operations. This section presents information for analysts to develop some of those products that can enhance the probability of successful operations.

a. MDCI analysts, interrogators, and CI agents maintain the C-HUMINT database. Using this database, they produce—

   (1) Time event charts.
   (2) Association matrices.
   (3) Activities matrices.
   (4) Link diagrams.
   (5) HUMINT communication diagrams.
   (6) HUMINT situation overlays.
   (7) HUMINT-related portions of the threat assessment.
   (8) CI target lists.

b. The analytical techniques used in HUMINT analysis enable the analyst to visualize large amounts of data in graphic form. We emphasize, however, that these analytical techniques are only tools used to arrive at a logical and correct solution to a complex problem; the techniques themselves are not the solution.

c. There are three basic techniques (tools) used as aids in analyzing HUMINT-related problems. These techniques—time event charting, matrix manipulation, and link diagraming—used together, are critical to the process of transforming diverse and incomplete bits of

seemingly unrelated data into an understandable overview of an exceedingly complex situation.

(1) Time event charting.

    (a) The time event chart shown in Figure A-VI-1,
is a chronological record of individual or group
activities designed to store and display large
amounts of information in as little space as
possible. This tool is easy to prepare, understand, and use. Symbols used in time event
charting are very simple. Analysts use triangles
to show the beginning and end of the chart.
They also use triangles within the chart to
show shifts in method of operation or change
in ideology. Rectangles or diamonds are used
to indicate significant events or activities.

    (b) Analysts can highlight particularly noteworthy or
important events by drawing an "X" through the
event symbol (rectangle or diamond). Each of these
symbols contains a chronological number (event

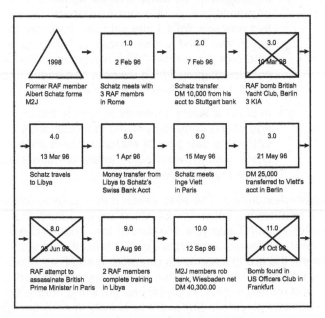

Figure A-VI-1. Sample time event chart.

143

number), date (day, month, and year of event), and may contain a file reference number. The incident description is a very brief explanation of the incident, and may include team size, type of incident or activity, place and method of operation, and duration of incident. Time flow is indicated by arrows.

(c) Analysts also use a variety of symbols such as parallelograms and pentagons, and others, to show different types of events and activities. Using these symbols and brief descriptions, the MDCI analyst can analyze the group's activities, transitions, trends, and operational patterns. Time event charts are excellent briefing aids as well as flexible analytical tools.

(2) Matrix manipulation.

(a) Construction of a matrix is the easiest and simplest way to show relationships between similar or dissimilar associated items. The "items" can be anything relevant to the situation under investigation: persons, events, addressees, organizations, or telephone numbers. During this process, MDCI analysts use matrices to determine "who knows whom" or "who has been where or done what." This results in a clear and concise display which viewers can easily understand simply by looking at the matrix.

(b) In general terms, matrices resemble the mileage charts commonly found in a road atlas. Within the category of matrices, there are two types used in investigative analysis—association matrix and activities matrix.

i. Association matrix. The association matrix is used to show that a relationship between individuals exists. Within the realm of HUMINT analysis, the part of the problem deserving the most analytical effort is the group itself. Analysts examine the group's elements (members) and their relationships with other members, other groups and associated entities, and related events. Analysts can show the connections between key players in any event or activity in an association matrix shown in Figure A-VI-2. It shows associations within a

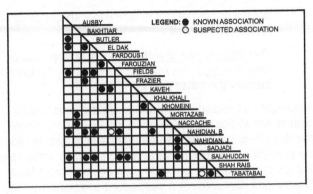

Figure A-VI-2. Sample association matrix.

group or similar activity, and is based on the assumption that people involved in a collective activity know one another.

(i) This type of matrix is constructed in the form of a right triangle having the same number of rows and columns. Analysts list personalities in exactly the same order along both the rows and columns to ensure that all possible associations are shown correctly. The purpose of the personality matrix is to show who knows whom. Analysts determine a known association by "direct contact" between individuals. They determine direct contact by a number of factors, including face-to-face meetings, confirmed telephonic conversation between known parties, and all members of a particular organizational cell.

NOTE: When a person of interest dies, a diamond is drawn next to his or her name on the matrix.

(ii) MDCI analysts indicate a known association between individuals on the matrix by a dot or filled-in circle. They consider suspected or "weak" associations between persons of interest to be associations which are possible or even probable, but cannot be confirmed using the above criteria. Examples of suspected associations include—

- A known party calling a known telephone number (the analyst knows to whom the telephone number is listed), but cannot determine with certainty who answered the call.

- The analyst can identify one party to a face-to-face meeting, but may be able to only tentatively identify the other party.

(iii) Weak or suspected associations on the personality matrix are indicated by an open circle. The rationale for depicting suspected associations is to get as close as possible to an objective analytic solution while staying as close as possible to known or confirmed facts. If analysts can confirm a suspected association, they can make the appropriate adjustment on the personality matrix.

(iv) A secondary reason for depicting suspected associations is that it may give the analyst a focus for tasking limited intelligence collection assets to confirm the suspected association. An important point to remember about using the personality matrix: it will show only that relationships exist; not the nature, degree, or frequency of those relationships.

ii. Activities matrix. The activities matrix is used to determine connectivity between individuals and any organization, event, entity, address, activity, or anything other than persons. Unlike the association matrix, the activities matrix is constructed in the form of a square or a rectangle as shown in Figure A-VI-3. It does not necessarily have the same number of rows and columns. The analyst can tailor rows or columns to fit the needs of the problem at hand or add them later as the problem expands in scope. The analyst determines the number of rows and columns by the needs of the problem and by the amount of information available.

(i) Analysts normally construct this matrix with personalities arranged in a vertical listing on the left side of the matrix; and activities, organizations, events, addresses, or any other common denominator arranged along the bottom of the matrix.

(ii) This matrix can store an incredible amount of information about a particular organization or group, and can build on information developed in the association

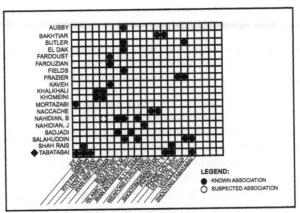

Figure A-VI-3. Sample activities matrix.

matrix. Starting with fragmentary information, the activities matrix can reveal an organization's—

- Membership.
- Organizational structure.
- Cell structures and size.
- Communications network.
- Support structure.
- Linkages with other organizations and entities.
- Group activities and operations.
- Organizational and national or international ties.

(iii) As with the association matrix, known association between persons and entities is indicated by a solid circle, and suspected associations by an open circle.

(iv) Analysts use matrices to present briefings, present evidence, or store information in a concise and understandable manner within a database. Matrices augment, but cannot replace, standard reporting procedures or standard database files. Using matrices, the analyst can—

- Pinpoint the optimal targets for further intelligence collection.
- Identify key personalities within an organization.

147

- Increase the analyst's understanding of an organization and its structure.

NOTE: The graphics involved in constructing the two
types of matrices differ slightly, but the principles are the
same.

(3) Link diagraming. The third analytical technique is
    link diagraming shown in Figure A-VI-4. Analysts use
    this technique to depict the more complex linkages
    between a large number of entities, be they per-
    sons, events, organizations, or almost anything else.
    Analysts use link analysis in a variety of complex
    investigative efforts including criminal investigations,
    terrorism, analysis, and even medical research.
    Several regional law enforcement training centers
    are currently teaching this method as a technique in

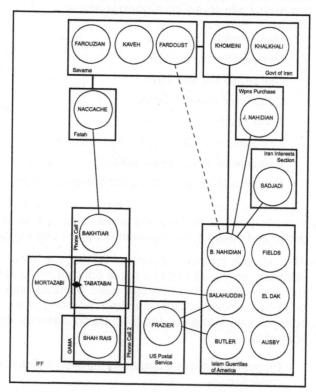

Figure A-VI-4. Sample link diagram.

148

combatting organized crime. The particular method discussed here is an adaptation especially useful in CI investigative analysis in general and terrorism analysis in particular.

a. The difference between matrices and link analysis is roughly the same as the difference between a mileage chart and a road map. The mileage chart shows the connections between cities using numbers to represent travel distances. The map uses symbols that represent cities, locations, and roads to show how two or more locations are linked to each other. Different symbols on the map have different meanings, and it is easy to display or discover the best route between two or more locations as well as identify obstacles such as unpaved roads or bodies of water.

b. The same is true with link analysis. Different symbols are used to identify different items. Analysts can easily and clearly display obstacles, indirect routes or connections, and suspected connections. In many cases, the viewer can work with and follow the picture easier than the matrix. Link analysis can present information in a manner that ensures clarity.

c. As with construction of association matrices, certain rules of graphics, symbology, and construction must be followed. Standardization is critical to ensure that everyone constructing, using, or reading a link diagram understands exactly what the diagram depicts. The standard rules follow:

i. Show persons as open circles with the name written inside the circle.

ii. Show persons known by more than one name (alias, also known as [AKA]) as overlapping circles with names in each circle.

iii. Show deceased persons as above, with a diamond next to the circle representing that person.

iv. Show nonpersonal entities (organizations, governments, events, locations) by squares or rectangles.

v. Show linkages or associations by lines: solid for confirmed and dotted for suspected.

vi. Show each person or nonpersonal entity only once in a link diagram.

d. Certain conventions must be followed. For the sake of clarity, analysts arrange circles and squares so that whenever possible, lines of connectivity do not cross. Often, particularly when dealing with a large or especially complex problem, it is difficult to construct a link diagram so that no connecting lines cross. Intersecting lines, however, muddle the drawing and reduce clarity. If lines must cross, show the cross-ing as a crossing, not as an intersection, in exactly the same manner as on an electrical schematic or diagram.

e. Link diagrams can show organizations, membership within the organization, action teams or cells, or partici-pants in an event. Since each individual depicted on a link diagram is shown only once, and some individuals may belong to more than one organization or take part in more than one event, squares or rectangles representing nonpersonal entities may overlap.

f. Construct the appropriate association matrices showing "who knows whom," "who participated in what," "who went where," and "who belongs to what group."

g. Draw information from the database and intelligence reports, and relationships from the matrices. Group persons into organizations or cells based on informa-tion about joint association, activities, or membership. Draw lines representing connections between individu-als, organizations, or activities to complete the diagram. You may have to rearrange the diagram to comply with procedural guidelines, such as crossed lines of connec-tivity. The finished product will clearly display linkages between individuals, organizations, and other groupings.

h. When you finish the matrices and link diagram, make recommendations about the group's structure. Identify areas for further intelligence collection targeting. Task intelligence assets to confirm suspected linkages and identify key personalities for exploitation or neutraliza-tion. The combination of matrix manipulation and the link diagram present, in effect, a graphic depiction of an extremely complex threat situation in a clear and concise picture.

i. Overlapping organizations.

    i. There is more to overlapping organizations than is immediately obvious. At first glance, the over-lap indicates only that an individual may belong to more than one organization or has taken part in multiple activities. Further study and analysis would reveal connections between organizations, con-nections between events, or connections between organizations and events.

    ii. When, as is often the case, an organization or incident shown in a link diagram contains the names of more than one individual, it is not neces-sary to draw a solid line between those individuals to indicate connectivity. We assume that individual members of the same group or participants in the same activity know each other, and the connection between them is therefore implied.

j. A final set of rules for link diagrams concerns con-nectivity between individuals who are not members of an organization or participants in an activity, but who are somehow connected to the group or activity. Two possibilities exist: The individual knows a mem-ber or members of the organization but is not directly connected with the organization itself. The person is somehow connected with the organization or activity but cannot be directly linked with any particular member of that organization or activity. In the first case, draw the connectivity line between the circle representing the individual and the circle representing the person within the organization or activity.

k. If you keep in mind the preceding outline of principles and rules, you can construct a link diagram effectively. Because this is a rather complex form of analytical graphic display to construct, it may prove difficult at first and require a little extra time and effort. The payoff, however, is the powerful impact of the results, which are well worth the extra effort.

Section VII
Personnel Security Investigations
to
Appendix A
Counter-Human Intelligence Techniques
and Procedures

A-VII-1. Personnel Security Investigations.

CI agents conduct PSIs on individuals requiring access to classified information. DIS Manual 20-1-M contains personnel security investigative requirements, types and scope of investigations, and the criteria for each component of a PSI. It also contains the methods and procedures governing the conduct of PSIs. In CONUS, DIS conducts PSIs; and OCONUS, the military services conduct PSIs on behalf of DIS.

a.   There are several types of PSIs. Each type provides the individual with a different level of access to classified information. The types of PSIs are national agency checks (NACs), single scope background investigations (SSBIs), and LAA.

   (1) NAC. An NAC consists, as a minimum, of a check of the Defense Central Index of Investigations (DCII) and the FBI. The FBI check is a review of files for information of a security nature which is developed during applicant-type investigations. It also includes a technical fingerprint search (classification of SUBJECT's fingerprints and comparison with fingerprints on file). If the fingerprint card is not classifiable, a "name check only" is automatically conducted. Office of Personnel Management, INS, State Department, CIA, and other federal agencies also may be checked, depending on the case. An NAC is the minimum investigative requirement for a final SECRET clearance for military personnel. It may be used as a basis for an interim TOP SECRET clearance based on simultaneous submission of a request for a background investigation.

(2) SSBI. An SSBI consists of a records review NAC, interviews with sources of information, and a subject interview; the subject interview being the principal component. The SSBI covers the most recent 10 years of an individual's life or 18th birthday, whichever is shorter, provided the last two years are covered. No investigation is made for the period before the individual's 16th birthday. The SSBI includes local agency checks, interviews of developed character references, employment references with employment records checks, education records checks and interviews, interviews of neighbors at previous residences, credit checks, citizenship verification, plus selected follow-up interviews as required to resolve unfavorable or questionable information. An SSBI is the minimum investigative requirement for granting a final TOP SECRET security clearance or for participation in certain special programs.

(3) LAA. An LAA is the formal authority granted to non-US citizens to have access to specifically prescribed and limited US classified defense information and materials. In each case, an investigation equivalent to the SSBI in scope must be completed with favorable results. A polygraph may be used to compensate for those portions of the SSBI which cannot be accomplished due to geographic or procedural limitations. An LAA may remain in effect for a maximum of five years before reinvestigation.

b. DIS Manual 20-1-M contains a full explanation of PSI requirements and criteria.

A-VII-2. PSI Reference Interview.

Before conducting a PSI Reference (Source) Interview, the CI agent must carefully examine all available background information on the case without exceeding the agent's standing investigative authority, which allows checks of sources only when identity or reliability is questioned. The SUBJECT of the investigation may have submitted a DD Form 398 which is often the initial source of leads. It may indicate

the relationship between the listed references and the subject. It may also assist in creating a mental picture of the
SUBJECT—an invaluable aid in formulating a line of questioning for an interview.

a. Source interviews may be t asked by some form of
    lead sheet, which normally contains limited back-
    ground information. Leads may also be developed
    through previous investigative activity. Unit files, local
    and Federal law enforcement agency files, telephone
    books, and city directories are all sources of informa-
    tion on both the SUBJECT and potential sources. It
    is desirable and necessary in critical cases, where
    the credibility of an interviewee is questionable, to
    learn something about the Source. A telephone call to
    arrange an appointment with a prospective Source is
    a courtesy that is often helpful to the investigator. The
    final preparatory step is to form a tentative plan for
    questioning the Source.

b. The approach to a PSI Reference (Source) Interview is
    simply an application of the social code of polite behav-
    ior, together with certain investigative requirements. The
    CI agent must—

    (1) Determine that the right person has been con-
        tacted, using the full name of the interviewee or
        Source to prevent any possibility of error.

    (2) Identify the SUBJECT of the investigation and find
        out if the Source is or was acquainted with the
        SUBJECT.

    (3) Present credentials to the Source for inspection,
        even if the individual was previously contacted by
        telephone.

    (4) Ensure, to the extent possible, that the interview
        will not be interrupted or overheard. Emphasize the
        US Army policy limiting dissemination of details of
        an investigation.

    (5) Explain the purpose of the interview to the Source.
        The CI agent should emphasize the importance the
        Source's knowledge may have to the investigation.

Some people are inclined to look with suspicion at investigators and are reluctant to give information. A patient explanation of the purpose of the interview, and the important part the Source's cooperation may play, may be sufficient to allay this general suspicion and forestall any reluctance to provide information.

(6) Inform the Source that the interview and the matters discussed are regarded by the US Army as official US Army business and should not be discussed with other persons, especially the SUBJECT of the investigation.

(7) Advise the Source of the Privacy Act of 1974. Ask the Source if there are any objections to the release of the interviewee's name as the source of the information. This advisement should be given at the termination of the interview to both US citizens and resident aliens, unless Source raised the question of disclosure earlier in the interview. OCONUS interviews of foreign nationals, who are not US resident aliens, do not require a Privacy Act advisement.

(8) Establish rapport with the Source before beginning the interview. Proper rapport creates a mutual understanding between the parties of the interview. Professional appearance, a pleasant voice, a courteous demeanor, and a confident manner are all important. The burden for maintaining rapport throughout an interview rests with the CI agent. An interview normally takes place in the Source's home or place of work, where the individual is under no official compulsion to furnish the information sought. Topics of mutual interest should be used to help establish rapport, but caution must be exercised to keep the interview from becoming a casual conversation.

c. Once a Source is willing to cooperate, begin the questioning to establish the period of association with the SUBJECT and the extent of personal knowledge of the SUBJECT.

(1) The period of association consists of—

    (a) When the Source and the SUBJECT first met and under what circumstances.

    (b) When they last met and under what circumstances.

    (c) All periods of association.

    (d) Types of association, such as friends, coworkers, or both.

    (e) Frequency of contact, for both social and professional association.

    (f) Breaks in contact for periods of 30 days or more.

    (g) Whether there has been any form of communication between them since their last contact.

(2) This information also aids in developing systematic questions on the loyalty, integrity, discretion, and moral character of the SUBJECT. It is important that the CI agent lay a framework for the interview, through a thorough understanding of the association between the Source and the SUBJECT. In addition, the association area may disclose leads which may be exploited later.

d. The CI agent must exploit all aspects of the SUBJECT's background. The CI agent must be constantly alert for leads to other persons not listed by the SUBJECT as references. If the Source is or was an employee or coworker of the SUBJECT, attention should focus on the efficiency, initiative, and ability of the SUBJECT to get along with fellow workers and subordinates and on the SUBJECT's honesty, reliability, and general character. If the Source is or was a neighbor of the SUBJECT, the CI agent should discuss the SUBJECT's general reputation, family, leisure time activities, morals, and personal habits. Concentration on some points does not imply exclusion of others.

e. The CI agent must seek information which will assist in establishing the SUBJECTs loyalty, trustworthiness, and suitability. Figure A-VII-1 shows the general area of interest for an interview and a basis for discussion.

f. Avoid questions concerning religious beliefs, racial matters, politics, labor affiliations, or personal and domestic matters, unless absolutely essential to the investigation; such questions not relevant to the purpose of the interview constitute unnecessary and unwarranted invasion of the SUBJECT's privacy.

g. The CI agent, as a representative of MI, should be professional in manner and efficient in the execution of duty. The agent must be receptive and flexible.

  (1) The agent should dress either in civilian clothes or uniform according to assigned duty and mission. The agent pinpoints specific information desired and avoids general questions. When the Source states, for example, that the SUBJECT is an indiscreet person, request that the Source cite specific examples to support this opinion. If the Source claims the SUBJECT is a drunkard, the individual's definition of the term should be clearly established and specific details obtained.

  (2) Analyze each phase of the SUBJECT's background point-by-point.

  (3) If the Source presents information in a haphazard manner, the CI agent should guide the discussion into a logical pattern.

A-VII-3. PSI SUBJECT Interview.

a. CI agents conduct SUBJECT Interviews when tasked by DIS in the lead sheet.

b. If the SUBJECT is suspected of violating the law, the CI agent must advise the individual of rights under the provisions of the Fifth Amendment to the US Constitution or Article 31, UCMJ, as appropriate. The CI agent must remember that the SUBJECT has the right to legal advice at any time before, during, or after the interview. Note that there are no noncustodial interviews involving a military member according to Article 31, UCMJ.

(1) Birth.	SUBJECT'S date and place of birth.
(2) Education.	Names and addresses of institutions, dates of attendance. academic records and degrees received.
(3) Employment.	Names and addresses of employers and dates of employment, names of immediate superiors and coworkers, nature of duties, quality of performance, and reason for termination.
(4) Technical skills.	Education and circumstances of development.
(5) Interests.	Hobbies and avocations.
(6) Foreign activity.	Foreign countries traveled to for personal purposes; friends and relatives who reside in a foreign country, foreign friends, foreign relatives, foreign associates, business or financial interests in foreign countries, correspondence with or visits from persons residing in foreign countries, contact with foreign official representatives or embassies and membership in foreign organizations.
(7) Mental and emotional stability.	General alertness, natural inclination, and idiosyncrasies. Treatment for mental or emotional disorders.
(8) Moral character.	Personal habits, particularly virtues and faults, illegal use of narcotics, and excessive consumption of alcohol.
(9) Loyalty.	Belief in and adherence to the US Constitutional form of government and to the laws of the nation. Reaction to ideologies which are hostile to those of the US Constitutional form of government.
(10) Integrity.	Uprightness of moral character and strength of convictions. Honesty.
(11) Discretion.	Speech and behavior, judgment and self-control. Respectful of property.
(12) Reputation.	Social and professional.
(13) Records.	Adverse involvement with law enforcement authorities, both civilian and military. Receipt of courts-martial or nonjudicial punishment.

Figure A-VII-1. Areas of interest for an interview.

158

(14) Finances.	Financial stability and reliability. Evidence of excessive debt or credit problems. Credit bureaus. Evidence of living beyond ones means.
(15) Family background.	Origin of parents, relatives abroad, residences, and citizenship.
(16) Association.	Friends, business or other associates, and favorite haunts.
(17) Organizations.	Membership, active participation, position, professional societies, character of organization, financial contributions, and awareness of aims of organization.
(18) Leads.	Names and addresses of persons acquainted with various phases of the SUBJECT's background and sources of information not listed by subject as references.
19) Recommendations. (PSIs and as required)	Interviewee's overall opinion of the SUBJECT's qualifications for a position of trust and responsibility. Stress that the recommendation is based on the period of association or a prior period of association, if there is a break in contact between the Source and the SUBJECT.

Figure A-VII-1. Areas of interest for an interview (continued).

c. The CI agent should not raise questions concerning religious beliefs, racial matters, politics, labor affiliations, or personal and domestic matters unless directly related to the investigation. The CI agent should phrase and time such questions so as to clearly establish the fact that they are relevant to the investigation.

d. To prepare for a SUBJECT Interview, the CI agent—

(1) Contacts the SUBJECT and informs the individual of the reason for the interview. In most cases, the matters to be discussed will not come as a surprise to the SUBJECT. The CI agent tells the SUBJECT that the interview gives the individual an opportunity to explain, refute, or mitigate, questionable or misleading information, and to provide information

not otherwise obtainable. If the SUBJECT is willing to be interviewed, the CI agent arranges the time, date, and place for the interview. If the SUBJECT refuses to be interviewed or to answer questions, an official record should be made of the refusal. Provide a brief advisement to the SUBJECT that failure to provide information may adversely affect the processing of the SUBJECT's security clearance.

(2) Gathers all lead information before the interview. Carefully reviews preplanned questions for each interview so that only information specifically authorized by the control office is released to the SUBJECT during the interview.

(3) Ensures that the SUBJECT understands that upon request, and if the individual is called to appear before a field board of inquiry or a civilian security hearing board, the SUBJECT will be provided a copy of any statement provided during the inter-view. However, the restrictions in AR 380-5 on the release of classified information apply.

(4) Ensures the SUBJECT's copy will not bear a protective marking but will contain a statement substantially as follows: "A copy of (describe) is furnished at your request. The official copies of this document will be protected to safeguard your confidence and will be used for official purposes only."

e. In accordance with the Privacy Act of 1974, whenever a CI agent interviews a SUBJECT, the agent must give the SUBJECT a four-point Privacy Act Advisement. In cases where an advisement of rights is required, the CI agent should provide the SUBJECT with the Privacy Act Advisement statement before the SUBJECT is advised of individual rights under Article 31, UCMJ, or the Fifth Amendment to the US Constitution.

(1) The CI agent should provide the SUBJECT with two copies of the advisement statement. One copy is for the individual's retention, if desired. The CI agent will request the SUBJECT sign the other

copy and return it. Privacy Act of 1974 statements are retained but not attached as part of the report.

(2) DIS Manual 20-1-M covers Privacy Act Advisement procedures during the conduct of a PSI under the control of DIS.

(3) The CI agent will verbally inform the SUBJECT that the Privacy Act of 1974 requires that each individual asked to provide personal information be advised of the following four points:

    (a) Authority by which the information is being collected.

    (b) Principal purpose for which the information will be used.

    (c) Routine uses of the information.

    (d) Voluntary nature of disclosing information.

(4) Before highlighting the four points, the CI agent should allow sufficient time for the SUBJECT to read the advisement statement.

(5) After highlighting the four points, ask the individual to sign one copy before beginning the interview.

f.   During conduct of SUBJECT Interview, the SUBJECT perceives the CI agent as being a representative of the US Army. As such, the SUBJECT will regard the CI agent's every statement, question, or contact as part of the official proceeding, whether so intended or not. During the interview, the CI agent will—

(1) Make no off-the-record or unofficial remarks in the interview, nor any promises or commitments to the SUBJECT which are beyond the CI agent's legal authority to fulfill.

(2) Avoid statements or representations which may be construed as opinion or advice to the SUBJECT about past, present, or future actions. CI agents should not argue with the SUBJECT or express personal viewpoints on any matter.

(3) Obtain permission from the SUBJECT if a tape or other recorder is to be used during the interview.

DIS Manual 20-1-M does not require the use of a tape or other recording device. Take the following actions in the sequence listed. If not recording, omit portions that pertain to the recorder.

(a) Dictate identifying data into a tape recorder before the SUBJECT's arrival. Turn off the machine.

(b) Visually identify the SUBJECT; identify yourself and present credentials; and positively identify the SUBJECT through the use of a pictured ID card, recording all pertinent information from the ID card.

(c) Explain the general purpose and confidential nature of the interview.

(d) Obtain permission to record the interview. Explain that it will facilitate the preparation of a written transcript of the interview, which the SUBJECT will have an opportunity to review, correct, and sign under oath.

(e) Turn on the tape recorder. If the SUBJECT objects to the tape recorder, do not use it. Proceed, but take notes as accurately as possible, while maintaining close attention to the SUBJECT's verbal answers and physical reactions. A tape recording is an administrative convenience, but not having one will not unduly hamper taking the sworn statement and preparing the report.

(f) Administer a full explanation of rights (if required). Request the SUBJECT read and sign DA Form 3881 to acknowledge receipt of the explanation of rights; and to record the individual's decision to exercise or waive the right to remain silent and to consult counsel. If the individual exercises his or her right to silence or to consult counsel, the interview should terminate at this point.

(g) Advise the SUBJECT of the provisions of the Privacy Act of 1974. The CI agent will request the SUBJECT complete the Privacy Act of 1974 Advisement Statement.

(h) Explain to the SUBJECT the DA policy allowing SUBJECTs of investigations every reasonable opportunity to explain, refute, or mitigate information which is developed during the course of an investigation. Furthermore, explain that this is the individual's opportunity to provide whatever information the SUBJECT feels appropriate.

(i) Though not required by DIS Manual 20-1-M, the CI agent may administer the oath of truthfulness to the SUBJECT, following the explanation and acknowledgment of legal rights, but before asking any questions. An appropriate oath is: "Do you affirm that the statements you are about to make are the truth, the whole truth, and nothing but the truth?" Additional remarks such as "So help you God" are unnecessary and may be offensive. If the SUBJECT refuses to take an oath, ask why, then proceed.

(j) Ask the SUBJECT to state his or her name, rank, SSN, DPOB, unit of assignment, duty position, and residence address for the record.

(k) Conduct the interview using the questions from DIS Manual 20-1-M that cover the appropriate topics and prepared questions designed to elicit narrative answers. These prepared questions are only a guide and are not intended to be the only questions asked. The CI agent must fully develop all information provided by the SUBJECT. Record and report all answers accurately.

(l) Make arrangements for the SUBJECT to review and sign a typewritten sworn statement, before ending the interview.

(m) Thank the SUBJECT for cooperating and terminate the interview.

Section VIII
Counterintelligence Investigations
to
Appendix A
Counter-Human Intelligence Techniques
and Procedures

A-VIII-1. CI Investigations.

CI investigations are conducted when sabotage, espionage, spying, treason, sedition, or subversive activity is suspected or alleged. An investigation is initiated by a SAEDA report being submitted and a case being opened by the SCO and ACCO. The primary purpose of each investigation is to identify, neutralize, and exploit information of such a nature, form, and reliability, that may determine the extent and nature of action, if any, necessary to counteract the threat and enhance security.

a. The ACCO and individual SCOs control and direct investigations under the provisions of AR 381-20 and other applicable regulations.

b. The initial objective of investigations involving national security crimes is to determine the nature and extent of damage to national security. Our intent is to develop information of sufficient value to permit its use in the appropriate civil or military court or to initiate CE proce-dures. However, we should not limit such investigations to the production of evidence. The investigative reports should include all relevant and material information.

c. CI agents conducting investigations must have a thorough understanding of the objectives and opera-tions of foreign espionage, sabotage, and subversive organizations.

d. Investigations are generally incident investigations concerning acts or activities which are committed by, or involve, known or unknown persons or groups of per-sons. An incident case can involve one or several of the national security crimes: sabotage, espionage, spying, treason, sedition, or subversion. The definitions used in this appendix focus on elements of a crime as an aid to CI personnel in investigations.

A-VIII-2. Sabotage.

Sabotage is defined as an act, the intent of which is to damage the national defense structure. Intent in the sabotage statute means knowing that the result is practically certain to follow, regardless of any desire, purpose, or motive to achieve the result.

a. Because the first indication of sabotage normally will be the discovery of the injury, destruction, or defective production, most sabotage investigations involve an unknown person or persons.

b. We expect acts of sabotage, both in overseas AOs and in CONUS, to increase significantly in wartime. Sabotage is a particularly effective weapon of guerrilla and partisan groups, operating against logistic and communications installations in occupied hostile areas, and during insurgences. Trained saboteurs sponsored by hostile guerrilla, insurgent, or intelligence organizations may commit acts of sabotage. Individuals operating independently and motivated by revenge, hate, spite, or greed may also conduct sabotage. In internal defense or limited war situations where guerrilla forces are active, we must be careful to distinguish among those acts involving clandestine enemy agents, armed enemy units, or dissatisfied friendly personnel.

c. Normally, we categorize sabotage or suspected sabotage according to the means employed. The traditional types of sabotage are incendiary, explosive, and mechanical. In the future, nuclear and radiological, biological, chemical, magnetic, and electromagnetic means of sabotage will pose an even greater threat to military operations. FM 19-20 discusses the materials and devices used in these types of sabotage.

d. The US Army CIDC will assume the investigative lead for actual or suspected sabotage. The jurisdiction of

CI elements is limited to the CI aspects of known or suspected foreign-directed sabotage. CI elements monitor the CIDC investigation and attempt to ascertain the existence of hostile, enemy, or foreign government involvement or the intent of the sabotage. CI elements do not conduct their own separate investigation unless hostile or foreign government involvement is evident or suspected.

e. When CIDC determines that the saboteur is operating on behalf of a foreign power, national security objectives will take precedence over criminal objectives. In this case, CI takes the investigative lead. In situations where CIDC support is not available (such as OOTW), CI elements will conduct the investigation.

f. Sabotage investigations require immediate action. The possibility exists that the saboteur may still be near the scene, or that other military targets may require immediate or additional security protection to avoid or limit further damage. We must preserve and analyze the incident scene before evidence is altered or destroyed.

g. The investigation must proceed with objective and logical thoroughness. The standard investigative interrogatives apply:

(1) Who. Determine a list of probable suspects and establish a list of persons who witnessed or know about the act.

(2) What. Determine what military target was sabotaged and the degree of damage to the target (both monetary and operational).

(3) When. Establish the exact time when the act of sabotage was initiated and when it was discovered; confirm from as many sources as possible.

(4) Where. Determine the precise location of the target and its relation to surrounding activities.

(5) Why. Establish all possible reasons for the sabotage act through the investigation of suspects determined to have had motive, ability, and opportunity to accomplish the act.

(6) How. Establish the type of sabotage (such as incendiary, explosive, chemical) and determine the procedures and materials employed through investigation and technical examination and analysis.

h. Destruction of government property. When destruction of government property is involved, CIDC will initially investigate the incident. Upon determination of intent to sabotage, CI and CIDC personnel may conduct a joint investigation of the incident. CIDC will normally retain the investigative lead.

i. An outline of possible investigative actions which may be used to investigate alleged or suspected sabotage incidents follows:

(1) Obtain and analyze the details surrounding the initial reporting of the incident to the CIDC unit. Establish the identity of the person reporting the incident and the reasons for doing so. Determine the facts connected with the reported discovery of the sabotage and examine them for possible discrepancies.

(2) Examine the incident scene as quickly as possible. The CI agent will attempt to reach the scene before possible sources have dispersed and evidence has been disturbed.

(a) The CI agent helps MP personnel protect the scene from disruption. The MP will remove all unauthorized persons from the area, rope off the area as necessary, and post guards to deny entrance and prevent anything from being removed.

(b) Although CI agents should help MP investigators at the crime scene, they should not interfere with the crime scene investigation.

(c) The CI agent may help CIDC personnel process the crime scene, to include locating all possible sources for questioning. CI keeps sources separated only in the sense that CI identifies to the MP or CIDC which ones should be separated. The physical act of separating is an MP or CIDC job.

(3) Preserve the incident scene by taking notes, making detailed sketches, and taking pictures. Arrange for technical experts to help search the scene and collect and preserve physical evidence and obtain all possible clues. Arson specialists, explosives experts, or other types of technicians may be required. Take steps to prevent further damage to the target and to safeguard classified information or material.

(4) Interview sources and obtain sworn statements as soon as possible to reduce the possibility of forgetting details or comparing stories.

(5) Determine the necessary files to be checked. These will be based on examination of the incident scene and by source interviews. CI conducts such action only in coordination with CIDC. CIDC has the crime scene expertise and responsibility; CI has the modus operandi (MO) expertise to identify to the CIDC.

    (a) Files of particular importance may include—

        i.   Friendly unit MO files.

        ii. Partisan, guerrilla, or insurgent activity files.

        iii. Local police files on arsonists.

        iv. Local police MO files.

        v.   Foreign intelligence agency MO files.

        vi. Terrorist MO files.

        vii. Provost marshal files.

    (b) Files checks should include background information on sources and the person or persons who discovered or reported the sabotage.

j.   Study all available information such as evidence, technical and laboratory reports, statements of sources, and information from informants in preparation for interrogation of suspects. FM 19-20 contains investigative guidance particularly applicable to the investigation of incendiary sabotage.

168

A-VIII-3. Espionage.

Espionage, as defined in Article 106a, UCMJ, and Title 18, US Code, is the act, either directly or indirectly, of obtaining, delivering, transmitting, communicating, or receiving information in respect to national defense with the intent or reason to believe that the information may be used to the injury of the US or to the advantage of any foreign nation. The offense of espionage applies in time of peace or war. There are five elements of espionage.

They are contact or communication, collection, tradecraft, reward or motive, and travel. Any or all of these elements are identifiable in counterespionage investigations. If agents recognize the type of information they are trying to collect and analyze data in light of the elements, they have a better understanding of the case and can plan more appropriately.

a. Examples of the elements of espionage are that:

   (1) The accused communicated, delivered, or transmitted any document, writing, code book, signal book, sketch, photographic negative, blueprint, plan, map, model, note, instrument, appliance, or information relating to the national defense.

   (2) This matter was communicated, delivered, or transmitted to any foreign government, or to any faction or party or military or naval force within a foreign country, whether recognized or unrecognized by the US, or to any representative, officer, agent, employee, subject or citizen thereof, either directly or indirectly.

   (3) The accused did so with the intent or reason to believe that such matter would be used to the injury of the US or to the advantage of a foreign nation.

b. Article 106a, UCMJ, further specifies that the punishment for espionage shall be as a court-martial may direct, except that if the accused is found guilty of an offense that directly concerns nuclear weaponry, military spacecraft or satellites, early warning systems, or other means of defense or retaliation against large scale attack, war plans, COMINT or cryptographic information, or any other major weapon system or major element of defense strategy, the accused shall be punished by death or such other punishment as a court-martial may direct.

c. Most espionage investigations originate from reports of incidents involving unknown individuals or allegations regarding known perpetrators. CI agents also conduct investigations of incidents where the crime of espionage has not been formally established, but is only suspected (the theft of classified documents or material). Leads in espionage investigations may originate from a wide variety of sources, including—

(1) Reports from sensitive sources.

(2) Reports from other intelligence, security, and law enforcement agencies.

(3) Evidence of espionage discovered during inspections and surveys of classified document handling and storage procedures.

(4) Reports submitted by military and civilian personnel in accordance with AR 381-12.

(5) Evidence of espionage discovered during screening of refugees, line crossers, displaced persons, civilian internees, EPWs, defectors, and similar groups, in areas of armed conflict.

(6) Information developed during the course of routine PSIs.

d. No single set of investigative procedures is applicable to the conduct of espionage investigations. Espionage is made up of many different elements, and espionage investigations are not always aimed at the arrest and prosecution of the offender. Prosecution of espionage cases may be deferred to the Department of Justice (CONUS) or to the host country (OCONUS). CI agents responsible for such an investigation must have a thorough and up-to-date knowledge of espionage and counterespionage methods and procedures as discussed in FM 34-5 (S).

e. In espionage cases, use any or all of the investigative techniques and tools described in this manual and FM 34-5 (S).

(1) Determine what specific techniques to use on a case-by-case basis.

(2) Get the proper authorization to use investigative techniques.

(3) Conduct the investigation in accordance with current laws and regulations.

f. Records examinations may break the cover story of an espionage suspect. CI agents may use properly authorized physical or technical surveillance to obtain leads or evidence. They may use confidential or sensitive sources or undercover operations to locate and identify suspects. Investigative photography may provide evidence of an attempt to transmit national defense information to a foreign nation.

A-VIII-4. Spying.

The crime of spying, is defined in Article 106, UCMJ. Spying is strictly limited to a wartime military situation. This is governed by international law, particularly, the Geneva Conventions. Five basic elements are required to constitute the crime of spying:

a. It occurs only during time of war.

b. It is committed within a US military AO.

c. The accused must be caught while seeking information to communicate to the enemy.

d. The accused must have the intent of passing information to the enemy.

e. The accused must have been acting in a clandestine manner.

A-VIII-5. Treason.

The abuse of treason laws in British legal history led the framers of the US Constitution to include a limiting definition of treason. Article 3, Section III, of the US Constitution also imposes qualifications regarding the conviction of an individual for treason: "No person shall be convicted of treason unless on the testimony of two sources to the same overt actor on confession in open court."

a. Investigations in which treason is suspected or alleged are rare. Historically, most cases occur during wartime, or upon the conclusion of hostilities.

b. Allegations of treason may originate with liberated prisoners of war, interned US civilians, examination of captured enemy records, or interrogation of enemy military and civilian personnel.

c. Federal courts have recognized two distinct types of treason: levying war and aiding and comfort. Investigations will be conducted with a view toward establishing the elements of the particular type of treason.

    (1) The elements of levying war treason are the—

        (a) Accused owed allegiance to the US.

        (b) Accused organized a body of personnel into a military force.

        (c) Accused equipped these personnel with arms.

        (d) Accused made war or military movement with intent to overthrow the government. Levying war treason has not occurred frequently in American history.

    (2) Aiding and comfort treason applies to persons with dual citizenship in a wartime situation. The elements of aiding and comfort treason are the—

        (a) Accused owed allegiance to the US.

        (b) War was formally declared by Congress.

        (c) Accused gave aid and comfort to the enemy.

        (d) Accused gave such aid and comfort while adhering to the enemy's cause.

d. Records examination, interviews, and interrogations normally are the principal investigative techniques employed in treason investigations. The CI agent pays particular attention to the legal requirements governing the collection and preservation of evidence, especially the taking of statements from sources and suspects.

e. In many cases, the CI agent needs to consult regularly with legal authorities to ensure that the elements of proof are adequately established and that all applicable legal conditions and restrictions are met.

A-VIII-6. Aiding the Enemy.

Investigations conducted by CI agents to prove or disprove charges brought against an individual under Article 104, UCMJ, may sometimes be treated as treason cases.

a. Article 104, UCMJ, makes five distinct activities criminal. They are—

   (1) Aiding the enemy with arms, ammunition, supplies, money, or other things.

   (2) Attempting to aid the enemy by performing an overt act with intent to aid the enemy with certain arms, ammunition, supplies, money, or other things.

   (3) Without proper authority, harboring or protecting the enemy.

   (4) Without proper authority, giving intelligence to the enemy.

   (5) Without proper authority, communicating, corresponding, or holding intercourse with the enemy, either directly or indirectly.

b. CI agent personnel must prove that one or more of the prohibited acts occurred. The word "enemy" includes organized forces in time of war, any hostile body that our forces may be opposing, and includes civilians as well as members of military organizations. It is not restricted to enemy government personnel or members of its armed forces. "Enemy" included Communist forces in Vietnam and Korea.

A-VIII-7. Sedition.

Investigations regarding alleged or suspected sedition may be based on either the Federal Sedition Statute or the UCMJ. Leads or allegations which prompt a sedition investigation by control offices may come from many sources. However, they are most often based on information submitted by confidential sources which are contained in reports from other agencies or developed during the course of routine background investigations (BIs).

a. Investigations involving sedition may occur with equal frequency in either peacetime or periods of hostilities. Title 18, US Code, describes two types of sedition: seditious conspiracy and advocating the overthrow of the US Government.

   (1) Seditious conspiracy. Title 18, US Code, Section 2384, makes it a specific crime to conspire to overthrow the US Government.

      (a) Unlike the general conspiracy statute, which makes it a crime to conspire to commit any federal crime, the seditious conspiracy statute does not require the commission of an overt act towards fulfillment of the conspiracy's objective.

      (b) The crime of seditious conspiracy is complete when two or more persons have entered into agreement to overthrow the government or to prevent, hinder, or delay the execution of any law of the US.

      (c) Remember, seditious conspiracy is a conspiracy to actually overthrow and is distinct from a conspiracy to advocate overthrow.

   (2) Advocating overthrow. Title 18, US Code, Section 2385, (also known as the Smith Act), enumerates specific types of activity which, if done with the intent to cause the overthrow of the government by force or violence, constitute sedition. The prohibited acts are—

      (a) Advocating action and systematically teaching the duty or necessity of such overthrow.

      (b) Using words to incite imminent lawless action with the specific intent of overthrowing the US Government.

b. Court decisions on the advocacy of overthrow have established that the advocacy must be calculated to incite persons to take action toward the violent overthrow of the government. The mere advocacy or teaching the forcible overthrow of the government as an abstract principle, divorced from any effort to instigate action, does not constitute the crime of sedition under the Smith Act.

c. The requirement for the advocacy (incite persons to take action) is of particular significance to CI agents. In any case alleging violation of the Smith Act, they will direct considerable effort toward proving that the oral or written material involved intended to incite listeners or readers to take action.

A-VIII-8. Subversion.

Title 18, US Code, Section 2387 and 2388, and Article 94, UCMJ, make it criminal to advise or attempt to cause military members to mutiny. It makes it clearly illegal to try to undermine the loyalty, morale, or discipline of the military force of the US.

a. Many investigations of subversive activity are cases based on adverse loyalty information developed during routine—

(1) SSBI.

(2) SAEDA reports submitted by military or civilian personnel under AR 381-12.

(3) Reports from other intelligence and security agencies.

(4) Leads obtained directly from sources used in CI special operations.

b. Note that the terminology "criminal subversion," "subversive activity," "subversion," and "sedition against the military" are clues to the CI agent to turn to Title 18, US Code, Section 2387 and 2388, for detailed elements of the crime.

c. The objective of such an investigation may be to determine if there is a need for some type of administrative action; for example, removal of an individual from a sensitive assignment to protect the security of the military command.

A-VIII-9. Responsibilities and Controls.

a. The DCSINT, DA, exercises DA staff cognizance for CI investigations conducted by Army CI organizations. The DCSINT formulates policies for the conduct, management, direction, and control of CI investigations.

b. For the DCSINT, INSCOM maintains the ACCO for all Army CI investigations, special operations, and counterespionage projects. The ACCO exercises overall control and coordination of all Army CI investigations, and ultimate case control over all investigations.

c. The SCOs have been established in the theater support brigades of the 66th, 470th, and 500th MI Brigades, plus the 650th and 902d MI Groups. The SCOs have authority to initiate, direct, and terminate CI investigations in accordance with AR 381-20.

d. CI investigations must conform to laws and regulations. CI agents must report information accurately and completely. They maintain files and records to allow transfer of an investigation without loss of control or efficiency. Coordination with other CI or law enforcement organizations ensures that investigations are conducted as rapidly as possible. It also reduces duplication and assists in resolving conflicts when jurisdictional lines are unclear or overlap. CI investigations must be conducted as to avoid publicity. This is required to protect the rights of individuals and to preserve the security of investigative techniques.

A-VIII-10. Investigative Plan.

When required, CI personnel formulate an investigative plan at each operational level down to and including the individual CI agent. The SCO dictates which element will write the investigative plan. Normally, that element will be the lead investigative element.

a. Although this list is not all encompassing, an investigative plan should include as many of the following planning considerations as applicable:

(1) Purpose of the investigation.

(2) Phases or elements of the investigation which have been assigned.

(3) Whether the investigation is to be conducted overtly or discreetly.

(4) Priority and time permitted for completion.

(5) Special instructions or restrictions.

(6) Information from the unit or office files.

(7) Definition of the problem.

(8) Methods and sources used, to include surveillance and polygraph support.

(9) Coordination required.

b. We must update the investigative plan as new developments arise, including an ongoing analysis of the results.

A-VIII-11. Order of Investigation.

All investigations vary, and as such, all investigative plans will be different. The following actions are typically conducted during an investigation. Tailor investigative plans to each investigation. Investigative actions selected should be sequenced to ensure a swift and successful completion of the investigation.

a. Files and records checks for pertinent information.

b. Individual interviews for additional information and leads.

c. Exploitation of new leads and consolidation of all available data for analysis and planning a course of action (COA).

d. Surveillance, both physical and technical, of the SUBJECT.

e. Interrogation or interview of the SUBJECT to prove or disprove the allegations.

f. Polygraph examination.

A-VIII-12. Investigative Techniques.

a. The CI agent uses the following basic techniques in CI investigations and operations, as appropriate:

(1) Examine records to locate, gain access to, and extract pertinent data from diverse official and unofficial documents and records.

(2) Conduct interviews to obtain information. The type of interview conducted depends on the investigation.

(3) Use interrogation and elicitation techniques as additional methods to gather information.

(4) Conduct physical and technical surveillance to augment other investigative activities. See FM 34-5 (S) for a detailed explanation of surveillance operations, and AR 381-10, Procedures 5, 6, 8, and 9, for legal requirements pertaining to electronic surveillance, concealed monitoring, searches and examination of mail, and physical surveillance.

(5) Conduct search and seizure when necessary. Do not conduct searches unless directed by the SCO for the appropriate level commander. The CI agent may coordinate this activity with law enforcement agencies, depending on the nature of the investigation. The CI agent will consult with the supporting SJA to ensure that the requirements for establishing probable cause have been met. (Refer to AR 190-22, AR 195-5, and FM 19-20 for policy and techniques used in searches and seizures; and AR 381-10, Procedures 7 and 8 covering physical searches and search and examination of mail.)

A-VIII-13. CI (SAEDA) Walk-in Interview.

A walk-in is defined as an individual who seeks out US Army Intelligence (USAI) to volunteer information which is believed to be of intelligence value.

a. When interviewing such persons, the CI agent must consider the Source's motives for divulging information.

If the motive can be determined early in the interview, it can be valuable in evaluating the information supplied and in determining the nature and extent of the Source's knowledge and credibility. Motivation includes, but is not limited to—

(1) Ideology.

(2) Personal gain.

(3) Protection of self or family ties.

(4) Fear.

(5) Misunderstanding of the function and mission of USAI.

(6) Mental instability.

(7) Revenge.

(8) Compliance with AR 381-12.

(9) Awareness from attending SAEDA briefings.

b. The motivation may not always be known, and sources may not always be truthful about their motives. The primary concern of the CI agent is to obtain all information, both of intelligence and CI value. The CI agent should be alert to detect whether the Source provides leads for exploitation.

c. Walk-in sources who volunteer information, that USAI is not authorized by AR 381-10 to collect, will be referred to the proper local, state, military, or national authority. Information received from anonymous telephone callers or written messages will be handled the same way. If possible, fully identify all unsolicited sources of information.

(1) If the Source's information is of no interest to USAI, but may be of interest to another agency, refer the Source to the appropriate agency.

(2) If the Source refuses referral to the appropriate agency, the CI agent will fully debrief the Source concerning the information. The CI agent may furnish the information verbally to the appropriate agency; but in all cases, a written report will be

provided to the agency concerned containing the details given by the Source.

    (a) Provide information concerning the Source, except when the Source requests anonymity as a condition of providing information.

    (b) Records of referrals and reports of information volunteered by unsolicited sources may be retained indefinitely if the information volunteered indicated the existence of a threat to life and property or the violation of law. If not, retention is authorized for no longer than 90 days, unless further retention is required by law or by Army regulation.

d. The following steps, in the order given, are basic to Walk-in Interviews:

    (1) Put the Source at ease. After determining that a walk-in source has information of intelligence value, display the appropriate credentials.

        (a) Take the Source to a private place to conduct the interview. The CI agent's initial attitude frequently affects the success of the interview. The atmosphere should be pleasant and courteous, but professional. In accordance with the Privacy Act of 1974, the Source must be given a four point Privacy Act Advisement to include authority, principle purpose, routine uses, and voluntary and mandatory disclosure, prior to the CI agent obtaining the Source's personal information. Ask the Source for some form of identification, preferably one with a picture.

        (b) Record the pertinent data from the ID card and tactfully exit the room.

        (c) Using the identity information just obtained from the Source, check the office source or informant files to see what, if any, information on the Source is on file. Determine if the Source is listed as a crank, has a criminal record, or has reported information in the past, and if so, what was the validity and value of that information.

(d) If the Source is listed as a crank or a nuisance continue with the interview, but include this information in the appropriate memorandum.

(2) Let the Source tell the story. Suggest that the Source start the story from the beginning, using the Source's own words.

(a) Once started, let the Source talk without interruption. The CI agent should, however, guide the Source back if the Source strays from the basic story. From time to time, interject a word of acknowledgment or encouragement.

(b) At no time, give any indication of suspicion or disbelief, regardless of how incredulous the story may seem.

(c) While the Source gives an account for the first time, take minimal notes. Taking notes could distract the Source or the CI agent. Instead, pay close attention and make mental notes of the salient points as a guide for subsequent detailed interviewing.

(3) Review the story with the Source and take notes. Once the Source has finished telling the basic story, he or she generally will freely answer specific questions on the details. Being assured that the information will be kept in strict confidence, the Source will be less apprehensive of your note taking.

(a) Start at the beginning and proceed in a chronological order, using the salient features of the Source's account.

(b) Interview the Source concerning each detail in the account so that accurate, pertinent information is obtained, meticulously recorded, and that the basic interrogatives are answered for every situation. This step is crucial.

(4) Develop secondary information. The story and background frequently indicate that the Source may have further information of significant intelligence interest. Also develop this information fully.

(5) Terminate the interview. When you are certain that the Source has no further information, close the interview in a manner which leaves a favorable impression.

(a) At this point in the interview, ask the Source, point blank, what motivated him or her to come in and report the information; even if the Source volunteered a reason earlier in the interview.

(b) Obtain a sworn statement from the Source, regarding the information, if appropriate. It is best to have the Source write (or type) the statement.

(c) Advise the Source of the Privacy Act of 1974 and ask the Source for full name; rank (for military or DOD civilian personnel) or occupation for non-DOD personnel; duty position, unit of assignment (for military or DOD civilian personnel); SSN, DPOB (required for military or DOD civilian personnel, requested for non-DOD personnel); type of security clearance and level of access; date of last SAEDA briefing; and full current address.

(d) Determine who else knows about the incident or situation, either directly or indirectly.

(e) Advise US sources of the provisions of the Privacy Act of 1974, and determine the Source's desires regarding the release of the Source's identity.

(f) Determine the Source's willingness to be recontacted by a member of USAI or another agency should the need arise regarding the information provided. Obtain recontact information from the Source (work or residence).

(g) Have the Source execute a Disclosure Warning and attach the affirmation to the report as an exhibit.

(h) Express appreciation for the information received.

e. In preparing for and conducting a Walk-in Interview, the CI agent—

(1) Should adapt to the intellectual level of the source, exercise discretion, and avoid controversial discussions.

(2) Must obtain all names and whereabouts of other individuals who may directly or indirectly know the same information.

(3) Must remember security regulations and make no commitments which cannot be fulfilled.

A-VIII-14. CI (SAEDA) Source Interview.

A Source is a person who has direct personal knowledge concerning a factor series of facts. The important part of this statement is direct, personal knowledge of a fact. The CI agent is concerned with the person who gained this knowledge of an action or incident through one of the five senses. The walk-in volunteers information. The CI agent has to locate and convince a Source to talk and provide the desired information. The CI agent often must persuade a Source to answer questions.

a. The Source is important because this person can provide both direct evidence as well as data and leads. These may not be admissible in a legal proceeding, but may serve to aid further investigation.

b. The general principles observed in interviewing walk-ins also apply to Sources, but a few additional factors have a bearing on the questioning technique:

(1) The Source's reputation, social standing, profession, and the fact that the person's statements are recorded for possible use in court cause understandable psychological reactions. These psychological effects are occasionally discovered in the form of resistance to questioning or refusal to testify.

(2) Sources may not be able to keep personal prejudice from distorting the facts. Less conscientious persons may not even attempt objectivity.

(3) Individuals may not know they are capable of unwittingly distorting facts or that forgotten details

are being replaced with products of their imaginations. The longer the time lapse between incident and interview, the greater the possibilities of imagination altering facts.

c. There are certain circumstances and conditions which may be present and which may affect the evaluation of information received from a Source.

(1) Physical condition: The Source's physical condition at both the time of the incident and at the time of the interview must be taken into account. Knowledge comes to an individual through one or more of a person's senses. If there are any limiting factors to an individual's sensory ability, questions may arise concerning the individual's competence in observation.

(2) Mental condition: Of similar importance is the Source's mental competency. A Source must be able to perceive, comprehend, and report what has happened to be considered competent. If one of these factors is missing or diminished, the individual will not be a good Source.

(3) Age:

(a) Once a child has reached the "age of reason," the child's testimony may have the same weight as that of an adult. This age is when a child is able to differentiate between fantasy and fact, and report factually what has happened. There are individual differences in this development process; but normally, the age of reason is considered to be seven or eight years. The likelihood of children bearing false testimony in deliberate attempts to influence situations is relatively slight. On the other hand, the demand of logic does not hamper their vivid imaginations; they do exaggerate.

(b) On the other end of the spectrum, senility or mental competence may be a factor.

(c) In either case, it is not wise to make generalizations about age when evaluating sources.

However, it is important to be aware of the age factor and take that into consideration with each individual involved in a case.

(4) Objectivity: Probably one of the most important factors to be considered during the Source interview is objectivity.

    (a) Normally, people observe and remember only those things of interest to themselves. Strong personal prejudices influence the way people see and remember things.

    (b) When dealing with a Source, the CI agent must listen carefully to what is said to determine what these interests and prejudices are and what errors they may cause in the Source's responses, whether intentional or unintentional.

    (c) Most errors of this type are unintentional and due to faulty memory. Careful questioning can discover this.

(5) Time: The average person's testimony will be distorted in one way or another. The brain will attempt to fill in any gaps by drawing on previous experiences. The more time that elapses between the incident and the questioning, the more the Source's story may become distorted. Unfortunately, the CI agent may not have control over the time factor; but in any case, attempt to interview the Source as soon as possible. This will enhance the validity of the Source's statement.

d. The CI agent's task is further complicated because the agent may deal with sources whose attitudes require the CI agent to change technique. The following are types of cases which require special treatment:

(1) Some sources may flatly refuse to talk because of possible danger to themselves. The CI agent should attempt to elicit cooperation by appealing to the Source's sense of patriotism or civic responsibility, pointing out it is in the individual's personal interest to talk, or leading the person into a logical path of reasoning. Discussing the Privacy Act of 1974 early may ease the Source's fears.

(2) Some sources are eager to demonstrate their knowledge to prove to themselves they are indispensable members of society. They may be braggarts, they may talk too much, or they may be a "know it all." The CI agent must be patient and critically weigh everything said, separating truth from fiction by asking pertinent questions and analyzing information carefully by comparing it with other known facts.

(3) Some sources are timid while others suffer from emotional stress and nervous tension. There may be occasions when much will be gained by asking the Source questions when he or she is extremely vocal due to an emotional condition. After the Source calms down, the CI agent should ask the questions again.

(4) A habitual liar is obviously a poor source, but there are occasions when such a person is the only source of direct evidence against a SUBJECT. In such an instance, do not ignore this person because of this weakness. Habitual liars usually contradict themselves; and if one can be made to repeat the story often enough, the truth may emerge.

(5) If possible, question a drunken source on the spot. At the risk of being led through a conversational maze, the CI agent should talk with the Source and strive to extract disclosures which the Source might not make if sober. You may use these statements later as a basis for a formal interview or interrogation of the Source.

e. Ideally, the CI agent should question all sources at the scene of the incident and obtain their first-hand knowledge while events are still vividly impressed on their minds. If this is the case, the CI agent should make arrangements to reinterview sources in a more formal manner later. If the CI agent arrives at the scene of an incident long after it has occurred, there will be time lags, so the CI agent should obtain the names of all sources from officials on the scene. The CI agent should take the following actions:

(1) Whenever possible, the CI agent should attempt to find out as much as possible about the Source and

how he or she is related to the incident before the interview. Do this through routine records checks. The CI agent is trying to determine if there are any obvious factors which would preclude this individual from being able to serve as a source. Gathering this type of information continues during the interview. Before starting the actual interview, the CI agent should review ail known details of the case and prepare questions for the Source. These questions should include those which help establish the Source's capability.

(2) As with other types of interviews, there are certain things which must be established before the main part of the interview. The proper approach will help establish the necessary rapport with the Source. The CI agent—

   (a) Presents credentials and verbally verifies the name and rank of the individual. In many cases, you may ask for positive ID, such as a military ID card. The use of positive ID will be left to your discretion. However, do not allow the request for ID to interfere with the establishment of rapport with the Source.

   (b) Quickly verifies that the person was indeed at the location at the time of the incident. If not, ascertain if this person knows of anyone who was there. It would probably be appropriate to ask why someone would believe that this person was there. You need to advise the individual that it is not his activities that are under investigation, but that you are trying to obtain information regarding what he might have seen or heard. Do not become antagonistic; other sources may have been mistaken. Maintain rapport. The CI agent may need to talk to this individual at another time. If the person admits to being at the scene, proceed with the interview.

   (c) Ensures that the Source understands that the US Government considers your presence and all matters discussed during the interview to be

187

official in nature and not to be discussed with anyone.

(3) Allow the Source to tell the story. As with the Walk-in Interview, allow the Source to tell the story in narrative format all the way through.

   (a) Keep note taking to a minimum. Focus attention on the Source and listen to the story.

   (b) Some sources may be reluctant to talk and tell their story. Some people may wish not to become involved; others fear having to go to court or other legal proceedings and face cross-examination; and some may fear reprisals. These people may need reassurance before they will talk freely. The CI agent—

   i.   Should spend the time necessary with these people to establish rapport; and attempt to determine why they are reluctant to talk.

   ii. May promise a Source confidentiality as a condition of providing information. Remember, this is the only promise that can be made. Ensure that the Source understands exactly what this means.

   iii. Must NOT make any other promises to these people. In most cases, a simple appeal to duty or patriotism may motivate a reluctant Source after rapport has been established.

   (c) Again, once the Source is willing to talk, let the person tell the story all the way through.

(4) Ask clear, direct questions which elicit narrative responses. As with the Walk-in Interview, go back over the Source's story. The CI agent—

   (a) Must develop the complete story from this Source.

   (b) Must not assume what this Source means, based on previous interviews. Cover all information and incidents brought to your attention by the walk-in and any previous sources to ensure that this Source's observations are obtained.

(c) Develops any new information this Source identifies.

(d) Uses basic questioning techniques. The six basic interrogatives should form the basis for all questions. Ensure that the Source's responses are fully understood.

(e) Fully identifies leads mentioned by the Source, during the course of the interview. The CI agent must not assume any previous knowledge of any information provided by the Source.

(5) When terminating an interview, the CI agent—

(a) Obtains full identifying data on the Source, after developing the Source's story. Full identifying data includes: name (last name , first name, and middle initial); rank, branch (if applicable), SSN, DPOB, MOS or duty position, unit of assignment, residence address, expiration of term of service, anticipated TDY or permanent change of station (PCS) dates, security clearance, and access to classified information, including any special accesses.

(b) Ensures the Source prepares a handwritten statement before leaving the interview, if appropriate.

(c) Provides the Source with the appropriate Privacy Act Advisement. This is essentially the same as for other sources.

(d) Determines if the Source has discussed the incident with others, if so, obtain their identities.

(e) Determines if the Source is willing to be recontacted by USAI, if necessary. Obtain the Source's desires regarding recontact.

(f) Ensures the Source executes a Disclosure Warning. This will depend on the approach used in the interview.

      (g) Leaves the Source with a good frame of mind and thanks the individual for cooperating.

f.   Throughout the entire interview, from the first approach to termination, the CI agent must never express opinions concerning the case. Simply get the Source's story. Some individuals, upon hearing the CI agent's opinion, may change their story to what they think the CI agent wants to hear as opposed to what actually happened. Be precise in recording the information. The CI agent must accurately record what the Source said. Significant contradictions between this Source's story and those of other sources may be addressed in the "Agent's Notes" paragraph of the related memorandum.

g.   Agent's notes from counterespionage interviews must be maintained. The notes whether handwritten, audio, or video taped, are part of the case file, and can be subpoened. Notes are important if a case goes to trial. The judge could dismiss the case or tell the jury to disregard the agent's testimony if the agent can not support testimony with notes made at the time of interview.

A-VIII-15. CI (SAEDA) SUBJECT Interview

a.   When tasked by the SCO or ACCO, CI agents conduct an interview of the SUBJECT of an investigation.

b.   The CI agent must advise the individual of rights under the provisions of the Fifth Amendment to the US Constitution or Article 31, UCMJ, as appropriate if the SUBJECT is suspected of criminal wrongdoing. The CI agent must also remember that the SUBJECT has the right to legal advice at any time before, during, or after the interview.

c.   The CI agent should contact the SUBJECT and inform the individual of the reason for the interview, such as involvement in a security matter. The CI agent should tell the SUBJECT that the interview gives the individual an opportunity to refute, mitigate, or explain questionable or misleading information and to provide information not otherwise obtainable.

(1) If the SUBJECT is willing to be interviewed, the CI agent should arrange the time, date, and place for the interview.

(2) If the SUBJECT refuses to be interviewed or to answer questions, make an official record of the refusal.

d. Before the interview, the CI agent must gather all available information and pertinent leads. The CI agent—

(1) Carefully reviews preplanned questions for each interview so that only information specifically authorized by the control office is released to the SUBJECT during the interview.

(2) Conducts the interview in an area that is under the CI agent's control.

e. During conduct of SUBJECT Interview, the SUBJECT perceives the CI agent as a representative of the US Army. As such, the SUBJECT will regard the CI agent's every statement, question, or contact as part of the official proceeding, whether so intended or not. The CI agent—

(1) Will NOT make any off-the-record or unofficial remarks in the interview.

(2) Will NOT make any promises or commitments to the SUBJECT which are beyond the CI agent's legal authority to fulfill.

(3) Avoids statements or representations which may be construed as opinion or advice to the SUBJECT about past, present, or future actions. Does not argue with the SUBJECT or express personal viewpoints on any matter.

(4) Asks for a reason if the SUBJECT refuses to be interviewed, and records the SUBJECT's response. Does not exert any pressure in an attempt to change the SUBJECT's mind.

(5) Stops the questioning if the SUBJECT requests a lawyer. If the SUBJECT is subject to the UCMJ, assists the SUBJECT in contacting the Trial Defense Service, through the SJA, if necessary.

(6) Takes the following actions in the sequence listed, when conducting the interview.

   (a) Dictate identifying data into a tape recorder before the SUBJECT arrives. Turn off the machine. However, recording interviews is neither required nor desired.

   (b) Initially identify the SUBJECT; identify yourself and present credentials. Positively identify the SUBJECT through the use of a pictured ID card, recording all pertinent information from the ID card.

   (c) Explain the general purpose and confidential nature of the interview.

   (d) Obtain permission to record the interview. Explain that it will facilitate the preparation of a written transcript of the interview, which the SUBJECT will have an opportunity to review, correct, and sign under oath.

   (e) Turn on the tape recorder.

      i.   The CI agent should take notes during the interview, even if it is being electronically recorded.

      ii. If the SUBJECT objects to the tape recorder, turn it off. Continue the interview taking notes as accurately as possible, while maintaining close attention to the SUBJECT's verbal answers and physical reactions.

      iii. A tape recording is an administrative convenience, but not having one will not hamper taking the sworn statement and preparation of the Investigative Memorandum for Record (IMFR).

   (f) Administer a full explanation of rights (if required). Request the SUBJECT read and sign DA Form 3881 to acknowledge receipt of the explanation of rights and to record the individual's decision to exercise or waive the right to remain silent and to consult counsel. If the

SUBJECT does not waive his or her rights, termi-nate the interview.

(g) In accordance with the Privacy Act of 1974, whenever CI agents interview a SUBJECT, they must give the SUBJECT a four-point Privacy Act Advisement.

   i. The CI agent should provide the SUBJECT with two copies of the advisement state-ment. One copy is for the individual's retention, if desired; the other copy is for reporting purposes.

   ii. Before highlighting the four points, the CI agent should allow sufficient time for the SUBJECT to read the advisement statement.

   iii. The CI agent should then explain the points covered in the form by stating:

The US Army is authorized to conduct CI investigations in accord-ance with government directives; the result of the inquiry will enable DA officials to determine the nature and extent of action necessary to ensure the security of the Army; the information obtained from the individual will be furnished to authorized government officials; and the disclosure of personal information to the US Army is voluntary. However, failure to disclose necessary and relevant information which impedes the investigation may have an adverse impact on obtaining or keeping a security clearance or employment with the DA.

(h) Explain to the SUBJECT the DA policy of allow-ing SUBJECTS of investigations every reason-able opportunity to explain, refute, or mitigate information which is developed during an investi-gation. Furthermore, that this is the SUBJECT's opportunity to provide whatever information the SUBJECT feels is appropriate.

(i) Ask if the SUBJECT is willing to take an oath.

   i. If the SUBJECT is not, ask why not, and continue the interview.

193

ii. If the SUBJECT is willing to take an oath of truthfulness, an appropriate oath is: "Do you affirm that the statements you are about to make are the truth, the whole truth, and nothing but the truth?" Additional remarks such as "So help you God" are unnecessary and may be offensive.

(j) Ask the SUBJECT to state his or her name, rank, SSN, DPOB, unit of assignment, duty position, and residence address for the record.

(k) Conduct the interview using prepared questions designed to elicit narrative answers. These prepared questions are only a guide and are not intended to be the only questions asked. The CI agent must fully develop all information provided by the SUBJECT. Accurately record and report all answers.

(l) Only if tasked by the ACCO or SCO, determine the SUBJECT's willingness to submit to a polygraph examination. If a tape recorder is used, turn it off before asking this question.

(m) Obtain a sworn statement, preferably in the SUBJECT's own handwriting, before ending the interview. If illegible, prepare a typewritten sworn statement for the SUBJECT to review and sign. Include the original, handwritten statement as an attachment to the IMFR. Never destroy the original statement.

(n) Consistent with the offense the SUBJECT is under investigation for and the evidence available, arrangements for detention should be made prior to SUBJECT Interview.

(o) Remind the SUBJECT of the confidential and official nature of the interview and not to discuss it with anyone.

(p) Thank the SUBJECT for his or her cooperation and terminate the interview.

Appendix B
Counter-Signals Intelligence Techniques and Procedures

General

One of the most self-destructive aspects of any operation is complacency. We know we are the best and we are equipped and trained to employ the finest equipment available. Our problem is improper use of the resources given to us, thus providing our adversaries with the opportunity to maximize the effect of their often inferior equipment and techniques to support their actions against us. Overcoming complacency is part of the analyst's task in C-SIGINT. Knowing and understanding the adversary and his equipment, as well as the capabilities and limitations of our personnel and equipment, is the first step in countering hostile efforts.

Contents

This appendix provides the MDCI analyst with detailed, step by step procedures necessary to initiate a C-SIGINT support program or to fine-tune an existing one. This appendix also contains analytical techniques and procedures which include—

- Database.

- Threat Assessment.

- Vulnerability Assessment.

- Countermeasures Options Development.

- Countermeasures Evaluation.

It provides indepth coverage of the five-step C-SIGINT process

   (1) threat assessment,

   (2) vulnerability assessment,

(3) countermeasures options development,

(4) countermeasures implementation, and

(5) countermeasures evaluation

discussed in detail in Section II through Section V. Although you can apply the C-SIGINT process manually, automation is the standard tool for database manipulation and production of C-SIGINT. The All-Source Analysis System (ASAS) and compatible systems such as the Theater Rapid Response Intelligence Package (TRRIP) are the tools that make the national intelligence community become an intelligence asset responsive to the warfighter's requirements.

Section I
Database
to
Appendix B
Counter-Signals Intelligence Techniques and Procedures

B-I-1. General.

The MDCI analyst must establish a complete and accurate database before the C-SIGINT process can begin. Section I details the creation of the database necessary to support C-SIGINT. With an effective database, the analyst streamlines the entire five-step C-SIGINT process. The C-SIGINT portion of the CI database, hereafter referred to as the C-SIGINT database, organizes C-E information. The MDCI analyst implements the C-SIGINT database by automated procedures for ease in manipulating and maintaining information. He organizes the database to limit duplication of data and to assure the accuracy, quality, completeness, and integrity of the data.

B-I-2. Development.

a.   The MDCI analyst develops the database during the planning phase of an operation, before deployments begin. He conducts electronic preparation of the battle-field (EPB) for the command's AI. EPB is the systematic approach to determine, through SIGINT and electronic warfare support (ES), the echelons and disposition of the threat through the electromagnetic structure of the target. The MDCI analyst employs a five-step process in EPB.

(1)   Identification of expected electronic signatures.

(2)   Evaluation of the current electronic environment.

(3)   Comparison of expected situation with current situation.

(4)   Preparation of SIGINT/EW templates.

(5) Integration of SIGINT/EW templates with all-source intelligence.

b. For C-SIGINT purposes, the MDCI analyst employs EPB to identify echelons and disposition of friendly forces through the electromagnetic structure. The purpose of EPB in C-SIGINT analysis is to build the database in order to determine and analyze vulnerabilities to threat SIGINT and to reduce or eliminate those vulnerabilities. To perform EPB, MDCI analysts must determine friendly communications and noncommunications signatures which may be vulnerable to threat collection or EA. Upon deployment, MDCI analysts continuously update the database with information which could influence the development of countermeasures.

c. The MDCI analyst compiles the data for each step in the C-SIGINT process. Sources of the data include—

(1) Current messages, reports, plans, and orders.

(2) Interviews specific to a command.

(3) Army regulations and technical manuals.

(4) Reviews of tables of distribution and allowances (TDA) and tables of organization and equipment (TOEs).

d. No matter what storage means is used, the MDCI analyst organizes, manipulates, and maintains the data for immediate and subsequent use and review. Since the data are not useful without modification for analysis, the formats supporting analytic techniques, methods, and measurement are essential. Like the data, the formats must be easily accessible and complete. The database includes analytic support templates, maps, and formats.

B-I-3. Content.

a. Information in the database should include most OB factors and other pertinent information such as—

(1)   Composition.

(2)   Disposition.

(3)   Strength.

(4)   Tactics.

(5)   Training status.

(6)   C-E emitters or threat collectors.

(7)   EPB templates.

(8)   Situation overlays.

(9)   Intelligence summaries.

(10) Intelligence estimates.

b.   The MDCI analyst needs to crosswalk the C-SIGINT database with the rest of the CI database to ensure accuracy and currency of overall CI information. Because there are many databases to draw information from, the analyst can save considerable time and effort by being tied into the appropriate databases, and not redoing the work all over again. Analysts constantly review and update the database by analysis and provide reports to the commander.

B-I-4. Organization.

a.   The C-SIGINT database is organized to ease access to data. There are three rules for database organization and storage:

(1)   Store like data together if primarily used in a particular task or step.

(2)   Store data when first created, if they are shared or administrative data.

(3)   Store administrative and reference data separately from task support data.

b.   For data used in multiple steps or tasks or routinely updated, reference a data version. For example, analysts review the commander's operations plans (OPLANs) in the vulnerability assessment, and again in the countermeasure effectiveness evaluation. The second use of the OPLANs should reference the initial use and date in the vulnerability assessment. In addition to the shared resources of the CI database, the analyst maintains a note file for reminders, working aids, high priority items, procedures, and interpretations.

B-I-5. Collection.

a. To be an effective tool, the database requires full time dedicated personnel to maintain it. This ensures complete familiarity with friendly and threat systems, and the ability to compare threat to friendly data in a timely manner. Analysts seek the collection of data on two levels.

   (1) The first collection level, the technical data file, is a listing of the technical characteristics for the friendly command's emitters and the threat SIGINT/EW equipment. Sources for friendly technical information include the command's C-E officer, technical manuals, technical bulletins, system operators, and maintenance personnel. Analysts request information on threat systems, such as communications intelligence (COMINT) and electronic intelligence (ELINT) receivers and direction finding (DF) equipment and jammers, through the collection management element.

   (2) The second collection level is how the unit uses its specific equipment. The systems use file identifies how the friendly unit uses its emitters and how the threat uses its SIGINT/EW resources.

b. Where to begin and how to progress in the collection of data are simplified by establishing a prioritized database collection list. This list is based on how the threat might prioritize their SIGINT/EW targeting. Although adversary target priorities depend on the command level and may be altered as the tactical situation develops, they generally are—

(1) Artillery, rocket, and air force units that possess nuclear projectiles or missiles and their associated control systems.

(2) CPs, observation posts, communications centers (includes automated data processing), and radar stations.

(3) Field artillery, tactical air forces, and air defense units limited to conventional firepower.

(4) Reserve forces and logistic centers.

(5) Point targets that may jeopardize advancing threat forces.

c. The collection of friendly force information for technical data files requires research on all types of emitters. This must include more than just frequency modulation (FM) voice radios and ground surveillance radars. The various C-E emitters and ancillary equipment include but are not limited to the following:

(1) Single sideband voice radios.

(2) Facsimile.

(3) Multichannel transmitters.

(4) Antennas.

(5) Retransmission systems.

(6) Tactical satellite communications systems.

(7) Automatic data processing transmission lines.

(8) Radio and wire integration.

(9) COMSEC machine encryption systems.

(10) Fiber-optic cable systems.

(11) Telephone wire systems.

(12) Countermortar radar.

(13) Air defense artillery target tracking radar.

(14) Air defense artillery target acquisition radar.

(15) Aviation guidance beacon systems.

(16) Aviation identification, friend or foe (IFF).

(17) Aviation ground control approach radars.

(18) Balloon-launched weather data radiosondes.

(19) EW jamming equipment.

(20) Cellular phones.

B-I-6. Construction.

Analysis employing any means other than automated data processing systems is a waste of time and effort. The analyst can revert to stubby pencil mode in an emergency but it is only a temporary fix until automated data processing (ADP) is back on line. No longer is it necessary for the MDCI analyst to build a database from scratch. Adversary COMINT and ELINT information are already in a database, organized for use, and available. The analyst needs only to extract pertinent adversary information from the database to cover the friendly AO and AI. He then puts this information into a working file for his use. The analyst can add to, delete from, and manipulate the information in his file without affecting the database he drew information from. Once the analyst has extracted and copied the data needed, he creates a working file of friendly emitters for his use. The analyst now has two working files that are the basis for future analysis. The analyst begins working the data, performing the analysis to satisfy the commander's needs.

B-I-7. Use.

The MDCI analyst is responsible for the control (security and access), use, and development of reports from the database.

a.   Access to the C-SIGINT data is based on the "need to know."

b.   Reports are correlated data from the CI database. The database contains working aids to help the analyst present information. Automated databases provide considerable flexibility in structuring reports. Manual databases have less flexibility and require considerable time and attention to detail unrelated to the analytic process.

B-I-8. Maintaining the Database.

Several areas are particularly important for the MDCI analyst who must maintain the SIGINT database.

a.   The first is timely review and update of the data. The analyst must update the database regularly with the most recent, valid information available, including the results of each analysis.

b.   The second area of importance is data integrity. This includes maintaining the most current version of information, ensuring proper and valid data are available, and fulfilling priorities and administrative requirements.

c.   Finally, the MDCI analyst must ensure the database contents support the CI analysis process. Should requirements, policies, or procedures change, the analyst should review and modify the database.

Section II
Threat Assessment
to
Appendix B
Counter-Signals Intelligence Techniques and Procedures

B-II-1. General.

One of the key words in the definition of intelligence is enemy. We must know our adversary as well or better than we know ourselves. We need to know and understand the capabilities and limitations of the threat arrayed against us and how the threat can influence our operations and mission. Section II, the first step in the C-SIGINT process, provides extensive information on determining foreign technical and operational capabilities and intentions to detect, exploit, impair, or subvert the friendly C-E environment.

a. Threat assessment is the key in planning C-SIGINT operations. The subsequent steps are necessary only when a defined threat exists.

b. Threat assessment is a continuous activity. It takes place throughout the conflict spectrum. A specific threat assessment is required to support a specific operation or activity.

c. The MDCI analyst gathers and analyzes information. He interacts with staff elements and higher, lower, and adjacent units to obtain the necessary data and access to supportive databases. Command support and direction are essential to success in the threat assessment process.

d. The major information sources available to the MDCI analyst include—

 (1) Validated finished intelligence products.

 (2) Theater and national level SIGINT threat database.

(3) Previous tasking.

(4) Analyst experience.

(5) The CI database.

e. MDCI analysts must continue to refine this list and identify other sources of information that may be available for their particular AO.

B-II-2. Procedures.

There are six tasks associated with threat assessment. These tasks are presented in Figure B-II-1.

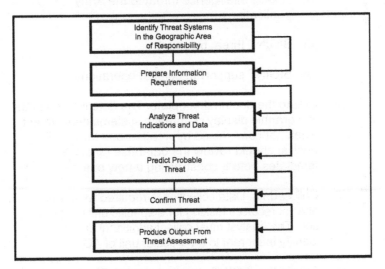

Figure B-II-1. Threat assessment process.

a. Identify threat systems in the geographic area of responsibility. This task provides the initial focus for the remaining threat assessment tasks. The primary objective of this task is to determine the specific threat faced by the supported commander. The MDCI analyst collects the required data to properly identify the threat. Additionally, the MDCI analyst must coordinate and request assistance from the collection management element. The procedures for identifying the threat systems follow:

(1) Identify the generic threat. The MDCI analyst enters the CI database and retrieves the most recent appropriate threat assessment. Analysts then review this data to determine what threat systems were known to be in their AO on the date of the assessment. Next, the analyst examines finished intelligence products published by national level agencies to obtain technical and operational data on the threat system. Some of the intelligence products include—

(a) ES and EA capability studies.

(b) Hostile intelligence threat to the Army publications.

(c) SIGINT threat by country.

(d) SIGINT support to combat operations.

(2) Create the doctrinal template. The doctrinal template is a graphic display of threat's systems deployment when not constrained by weather and terrain. The analyst should review the database for existing templates before constructing a new one.

(3) Collect data. Data collection is required when the analyst receives tasking for a specific unit or operation. The analyst must collect additional data to identify the threat to a particular unit or AO.

(4) Create the SIGINT situation overlay. The analyst reviews the collected data to determine—

(a) Technical and operational capabilities.

(b) Typical modes of operation.

(c) Current deployment.

(d) Probable tasking.

(e) Activities of the collectors of interest.

(5) Enter data. The analyst enters this data on the situation overlay.

(6) Summarize the data and identify the threat system. The MDCI analyst reviews the SIGINT situation overlay for patterns, electronic configurations, and threat command, control, and communications (C 3). The ACE has identified this information, which could help the analyst identify specific systems. A common approach is to pose and answer questions, such as—

(a) Is the threat system part of a larger system?

(b) What are the threat system's capabilities?

(c) How is the threat system doctrinally used?

(d) How does the threat system obtain information?

(e) How many collection systems were located?

(7) Request information. In some instances, sufficient information may not be available in the unit to make an accurate determination. For example, the type of equipment may be known but the technical characteristics of the system may not be available from local sources. If additional information is required, the MDCI analyst compiles the information needed and requests additional information from outside the unit.

b. Prepare information requirements.

(1) The MDCI analyst fills information shortfalls by requesting information from sources external to the unit. These external information sources are adjacent or higher echelons and national level assets. Each echelon satisfies a request with available data or organic assets, if possible. Requirements exceeding their organic capabilities are consolidated and forwarded to the next higher echelon as a request for information.

(2) Once a request reaches corps, the highest tactical echelon, the corps ACE provides the information

or passes the request to the theater MI brigade if it is beyond the capability of corps systems. This task requires the MDCI analyst to initiate a standard collection asset request format (SCARF) shown in Figure B-II-2 requesting information from higher or adjacent headquarters.

1.	Request Number:
2.	Originator priority:
3.	Activity and target type (area emitter, size, point or area, unit):
4.	BE number, ELINT notation, or case:
5.	Location (if known, last known):
6.	Duration:
	a. Start date and time:
	b. Frequency:
	c. Stop date and time:
	d. Latest acceptable date and time for information utility:
7.	Location accuracy:
	a. Required:
	b. Acceptable:
8.	PIR and IR:
9	Justification:
10	Remarks (to include disciplines and collectors recommended):

Figure B-II-2. Sample standard collection asset request format (SCARF).

(3) The SCARF is prepared in accordance with local SOP and the Joint Tactical Exploitation of National Systems (J-TENS) manual. At ECB units, this request is sent by a request for intelligence information (RII) using the US message text format (USMTF). The USMTF user's handbook provides instructions on preparing messages. The analyst forwards the request to the appropriate collection management section for action.

c. Analyze threat indicators and data.

(1) The MDCI analyst reviews, organizes, and evaluates key information components of the collected information. He evaluates the data looking for trends and

patterns of the threat system that will provide an estimate of capabilities and intentions. He focuses on each component of the collected information to determine if it reveals a tendency of the threat system to act or react in a particular manner. Additionally, the analyst evaluates the information for trends or characteristics that will aid in the ID and evaluation of the capabilities and intentions of the threat system. Additional support may be required from other staff elements.

(2) The procedures for analyzing threat indicators and data are to—

    (a) Compile and organize data. First, the analyst compiles and organizes the data that has been collected. He updates the database with new information and organizes the data into collector categories.

    (b) Review data. The analyst reviews the collected data to determine the ability of the threat systems to collect against a specific target.

    (c) Determine intentions. To determine the intentions of the threat system, the MDCI analyst poses the following questions and enters this information in the database:

        i.   What area will the threat system target?

        ii.   When will the targeting take place?

        iii.   Why is the targeting taking place?

        iv.   How will the threat system attempt to collect against the target?

        v.   How has the threat system been used in the past?

        vi.   What does threat doctrine suggest about probable threat?

        vii.   Does the threat system have a distinctive signature?

(3) Doctrinal templates are extracted from the database and compared to the SIGINT situation overlay. The analyst lists similarities between current and doctrinal deployments and selects the doctrinal template that has the greatest similarity to the current situation.

d. Predict probable threat.

(1) The MDCI analyst identifies the probable threat. He reviews all the information that has been collected and applies this information to the geographic AI and the capabilities and intentions of the threat system.

(2) The procedures for predicting the probable threat follow:

(a) Determine probable location. Use the SIGINT situation overlay and doctrinal templates to determine the location of the collectors. Overlay the doctrinal template over the situation overlay.

(b) Analyze terrain and weather effects. Integrate the terrain and weather data with the doctrinal template and the SIGINT situation overlay and create a situation template for the current environment. Terrain and weather conditions affect a threat system's ability to operate according to their doctrine. For example, a radio DF site must have a clear line of sight (LOS) on the emission of the target in order to gain an accurate bearing. Mountains, dense foliage, and water distort electronic emissions and impair a collector's ability to target. FM 34-130 provides information for military terrain and weather analysis.

(c) Update the SIGINT situation overlay. Place the symbols for the collectors on the doctrinal template that have not been confirmed on the SIGINT situation overlay as proposed locations.

e. Confirm threat. The MDCI analyst attempts to verify threat predictions. The procedures for confirming the threat follow:

(1) Validate existing data. Review current intelligence reports and assessments to determine if the information received from the original SCARF request and other information sources used in the assessment are valid. If there are indications that the capabilities or intentions of the threat system have changed, additional information may be required. This is determined by looking for information that could indicate a change in a collector's ability to collect against the command. For example, additional antennas have been added to the collector, or the collector has moved to provide for better targeting are indicators of a change in collection capabilities.

(2) Request additional information. If additional information is required, request this information by preparing a SCARF or request for information and forward it to the collection management section.

(3) Evaluate new information. If new information on the collector's intentions or capabilities is received, review this information to determine its impact on the original assessment, and update the situation overlay. If intentions and capabilities of the collector change, reevaluate the original threat prediction by following the tasks identified in previous sections.

f. Produce output from threat assessment. The MDCI analyst can present the threat assessment in briefings or reports. Portions of the threat assessment are included and presented in CI products.

Section III
Vulnerability Assessment
to
Appendix B
Counter-Signals Intelligence Techniques
and Procedures

B-III-1. General.

After examining our adversary's equipment, capabilities, and limitations, we now must examine our own unit to see how our adversary can affect us. Section III, the second step in the C-SIGINT process, details specific areas where a threat effort can be most damaging to the friendly force.

a.   The vulnerabilities are ranked according to the severity of their impact on the success of the friendly operation. The vulnerability assessment—

   (1) Examines the command's technical and operational C-E characteristics.

   (2) Collects and analyzes data to identify vulnerabilities.

   (3) Evaluates vulnerabilities in the context of the assessed threat.

   The MDCI analyst performs the primary data gathering and analysis required. Assistance by and coordination with the appropriate staff elements (intelligence, operations) is key to this process.

b.   Data gathering requires access to command personnel and to local databases. Data sources include—

   (1) Technical data on C-E inventories.

   (2) Doctrinal and SOP information.

   (3) Output from the threat assessment step.

(4) Command operational data.

(5) Essential elements of friendly information (EEFI).

(6) PIR and IR.

c. The database of friendly technical data is used throughout the vulnerability assessment process for key equipment information, mission data, and other supporting information.

d. Vulnerability assessment is comprised of ten tasks. The first three tasks are ongoing determinations of general susceptibilities. The next six are specific to the commander's guidance and involve determinations of specific vulnerabilities. The final task is the output. Vulnerability assessment tasks are shown in Figure B-III-1.

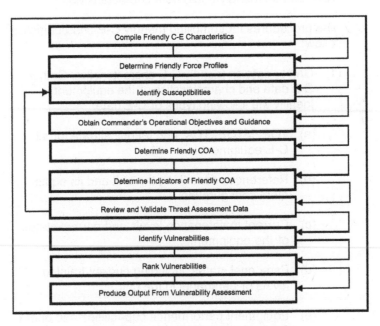

Figure B-III-1. Vulnerability assessment tasks.

B-III-2. Compile Friendly C-E Characteristics.

a. The MDCI analyst compiles friendly C-E characteristics. He collects and organizes unit C-E data and equipment characteristics for analysis. This analysis provides a baseline for analyzing friendly C-E equipment and operational susceptibilities to threat operations. The compilation of C-E characteristics is an ongoing process. Assistance from the command's C-E officer, property book officer, maintenance personnel, or system operators may be necessary.

b. The C-E data are a baseline for identifying friendly susceptibilities. A unit's equipment, personnel, and associated characteristics must be identified before the pattern and signature analysis can proceed. The MDCI analyst uses available databases to extract the TOE, modification table of organization and equipment (MTOE), and TDA data on friendly equipment characteristics.

c. The procedures for compiling friendly C-E characteristics follow:

   (1) Gather data on friendly C-E characteristics. Gather C-E data and characteristics of the equipment. Identify the following types of C-E data:

      (a) TOE, MTOE, TDA, and technical data for all C-E equipment in a unit.

      (b) References describing the unit and its equipment configuration.

      (c) Current maintenance levels and normal status of the equipment.

      (d) Personnel status, including current training levels of personnel in the unit.

      (e) Equipment performance capabilities and operational capabilities in all weather conditions, at night, over particular terrain, and toward the end of equipment maintenance schedules.

214

(f)   Equipment supply requirements.

(g)   Special combat support requirements.

(2)   Organize C-E data. The MDCI analyst organizes the information into a format useful for signature analysis. The data are organized by type of unit (if the support is multiunit), type of emitter, frequency range, number and type of vehicles or weapons which emit or carry emitters and the type of cluster. The electromagnetic overlay shown in Figure B-III-2 graphically depicts the friendly C-E equipment laydown on the battlefield.

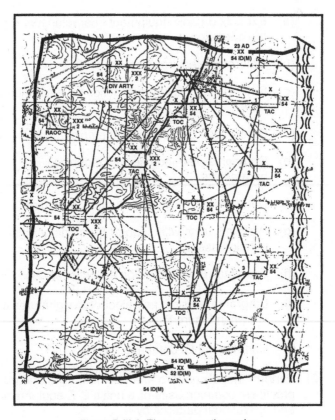

Figure B-III-2. Electromagnetic overlay.

215

B-III-3. Determine Friendly Force Profiles.

a. This task includes the analysis of signatures and patterns of the C-E equipment and a summary statement of the unit's C-E profile. A profile consists of the elements and standard actions, equipment, and details of a unit, the sum of signatures and patterns. SIGNATURES + PATTERNS = PROFILE

b. Procedures for determining the friendly force's profile follow:

(1) Analyze friendly force signatures. The MDCI analyst—

(a) Extracts organic equipment characteristics for the operation.

(b) Determines environmental effects.

(c) Determines C-E characteristics for each friendly COA.

(d) Determines C-E equipment employment.

(e) Compares planned use with technical parameters.

(f) Determines if further evaluation is required.

(g) Performs tests with support from unit or higher echelon assets.

(h) Evaluates the information collected above.

(i) Diagrams physical and electronic signatures as shown in Figure B-III-3.

(j) Updates the CI database.

(2) Perform friendly pattern analysis. Identify standard practices, common uses of a unit's C-E equipment, and operational patterns by—

Physical Signature: Additional vehicles camouflaged in the woods

    4 x M35A2
    1 x M104

Electronic Signature:

    AN/VRC-46 Div Cmd Net
    AN/GRC-106 DTOC SSB Net
    AN/GRC-142 Div Ops Intel RATT Net
    AN/TRC-145 MC to DIVARTY, DTAC CP, DISCOM
    AN/MRC-108B USAF Air Nets

Figure B-III-3 Physical and electronic signatures.

(a) Reviewing the database to obtain information that might provide the threat with critical data regarding unit type, disposition, activities, or capabilities.

(b) Extracting from the OPLAN and operations order (OPORD) particular means of communication, operational characteristics, and key and secondary nodes for communications support.

COMMAND AND CONTROL		
Physical Signatures	**Electronic Signatures**	**Pattern Data**
-Types of vehicles	-Types of emitters	-Timing of movement
-Number of vehicles	-Frequency range	-Mode of movement
-Distances to adjacent and higher echelon crops and HQ	-Signature type and range	-Collocated or nearby units
	-Emitter fingerprints	-Frequency of redeployments
		-Radio and radar net employment

OPERATIONS AND MANEUVER		
Physical Signatures	**Electronics Signatures**	**Pattern Data**
-Types of vehicles	-Types of emitters	-Timing of reconnaissance
-Number of vehicles	-Frequency range	-Mode of reconnaissance
-Distance to adjacent and higher echelon support elements	-Signature type and range	-Timing of movement
-Types of weapons systems	-Emitter fingerprints	-Type of movement
		-Mode of movement
		-Units involved
		-Mode and source of supply

Figure B-III-4. Friendly unit profile.

(c) Identifying specific patterns associated with types of operations.

(3) Correlate patterns and signature. In this subtask, compile the information from the signature and pattern analysis, which creates the profile. The analyst—

(a) Lists the signature and pattern data for particular types of C-E equipment.

(b) Matches signature with patterns to form the profile.

(c) Organizes data into types of C-E operations.

(d) Correlates signature and pattern data with past profiles to produce the current profile shown in Figure B-III-4.

(4) Produce unit profile. Patterns and signatures can change as commanders, staff, and operators change. Profile development must be an ongoing effort. To produce the unit profile, use the OPORD to obtain the past task organization and then select the areas of concern to that organization, that is, C 2 and maneuver.

B-III-4. Identify Susceptibilities.

a. The analyst determines how the profiles would appear to threat systems and which equipment or operations are susceptible. A susceptibility is defined as the degree to which a device, equipment, or weapon system is open to effective attack due to one or more inherent weaknesses. Any susceptibilities are potential vulnerabilities.

b. Information sources are of the following types:

(1) Current friendly C-E profile.

    (2) Historical profiles to compare with current profile.

    (3) Knowledge and experience from other analysts.

c. The procedures for identifying susceptibilities follow:

    (1) Identify weaknesses:

        (a) Review current profile and identify unique equipment or characteristics that the threat may use to determine intentions.

        (b) Review the CI database and compare historical profiles with current profile, noting correlations and deviations.

        (c) Plot friendly weaknesses to threat operations on the electronic order of battle (EOB) overlay shown in Figure B-III-5.

    (2) Categorize susceptibilities. Categorize susceptibilities to allow more specific analysis by equipment type, organization, and use. Do this—

        (a) By type (for example, equipment, operations, or both).

        (b) By activity (for example, logistic, C 3 , intelligence, operations, and administrative support).

        (c) According to resource requirements.

        (d) According to the length of time the susceptibility has existed.

        (e) According to scope (number of units or equipment types).

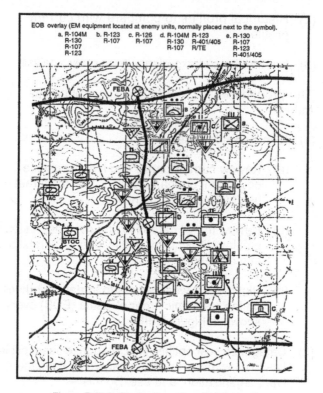

Figure B-III-5. Electronic order of battle overlay.

B-III-5. Obtain Commander's Operational Objectives and Guidance.

a.   The commander states his operational objectives for missions in OPLANs and OPORDs. The analyst uses this information to plan the most effective support for the commander and to identify the commander's prefer-ences for types of operations. The commander's opera-tional concept and EEFI shown in Figure B-III-6 are essential to the analysis of friendly COAs.

b.   This information enables the analyst to evaluate indi-cators of friendly COA in the context of what the commander considers essential to the success of the operation. Setting priorities for the vulnerabilities depends on the commander's operational concept. The primary information sources are—

(1) Concept of operation.

(2) OPORDs.

(3) OPLANs.

(4) EEFI.

Essential Elements of Friendly Information Statement

FRIENDLY SUPPORTED UNIT: 5th Inf Div

1. SUBORDINATE ELEMENT: HQ

2. LOCATION : 32 U N B51452035

3. OPERATIONAL OBJECTIVE:

Defend to PL Gray, counterattack at 300001Z Oct 96

4. EEFI:

a. SIGNIFICANT COMPROMISE

(1) Time of counterattack.

(2) Identification and location of HQ elements brigade and higher.

(3) Identification of attached units.

(4) Loss of C^3.

b. INSIGNIFICANT COMPROMISE: Identification of 5th Inf Div.

Figure B-III-6. Sample format for essential elements of friendly information.

B-III-6. Determine Friendly COA.

a. Based on the general description of the commander's objectives, the operations element plans locations and events. The analyst produces an overlay of the friendly force profile integrated with the commander's objectives.

b. The procedures for determining friendly COA follow:

(1) Identify COA. For each applicable level of command, identify friendly COA. At division level, for example, COA would include the following minimum information:

(a) Summary of operations.

(b) Corps and EAC support.

(2) Compare COA to specific EEFI. Review the COA for events or actions that could compromise the unit's mission by disclosing key EEFI. The review is summarized in an events list that describes a particular mission, COA, or event which may compromise the EEFI or the friendly intentions.

B-III-7. Determine Indicators of Friendly COA.

a. Indicators of friendly COA shown in Figure B-III-7 are those events and activities which, if known by the threat, would compromise a friendly COA.

b. The procedures for determining indicators of friendly COA follow:

(1) Identify the commander's preferences and perceptions about C-SIGINT operations. Seek information about the commander's style from sources such as previous concepts, plans, and orders, or interviews with subordinate commanders and staff officers.

(2) Integrate friendly profiles and COA. In the event planned location or movement data are not available, retrieve friendly operational overlays shown in Figure B-III-8 from the database. The overlays help identify friendly historical positions for the new COA. Figure B-III-9 depicts an example of an integration of a friendly force profile and COA. Integrate the friendly profile and COA by—

(a) Noting current position and expected COA.

(b) Identifying key C-E capabilities associated with the COA (for example, radio nets, types of radios, radar, teletypewriters).

(c) Noting past C-E operational patterns.

(d) Plotting critical C-E nodes, paths, or circuits.

(3) Determine standard C-E procedures for types of operations:

(a) Begin by using the commander's objectives to identify key operational constraints, that is,

222

nodes, paths, chokepoints, and standard C-E procedures followed during a particular COA. New or critical data, not previously included in the friendly profile and COA integration, are then added to the situation overlay.

(b) Also consider constraints and procedures while determining indicators. Document these as factors associated with those indicators in a format as in Figure B-III-7. After completing the review of existing data as obtained from the commander's objectives, determine what additional information is required.

INDICATORS OF FRIENDLY COAS 10-96			Version: OPORD Date : 10 Oct 96 Analyst: SSG L. Morrow	
Course of Action	Indicator Number			
	1 C-E Equipment	2 C-E Nets	3 Vehicle Movements	4
Defend to PL Gray	AN/TRC-145 AN/VRC-46 AN/GRC-142	Div MC Div Cmd Div Intel	Small vehicle movement around DTOC	
	FACTORS	FACTORS	FACTORS	FACTORS
	LOS required	LOS to brigade and DIVARTY	None	
Counterattack	AN/TRC-145 AN/MRC-108B	Div MC USAF	Pack up of vehicle USAF comm increase may be a tip off	
	FACTORS	FACTORS	FACTORS	FACTORS
	LOS required	LOS to brigade and DIVARTY	Must be in range for C^3	

Figure B-III-7. Indicators of friendly COA.

(4) Determine impact of weather and terrain. As the situation changes, the significance of particular nodes or paths may shift or additional nodes may become critical. Consider the following in determining the impact:

(a) Inclement weather.

(b) Night activity.

    (c) Terrain masking.

    (d) Poor C-E equipment maintenance.

    (e) Meaconing, intrusion, jamming, and interference (MIJI).

(5) Set priorities. Once the type of operation is determined, set priorities for the events, movements, and nodes by their overall importance to the operation.

(6) Identify critical C-E nodes—

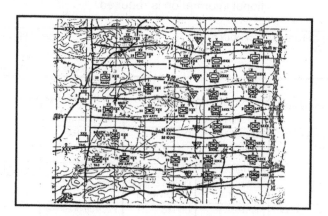

Figure B-III-8. Friendly operational overlay.

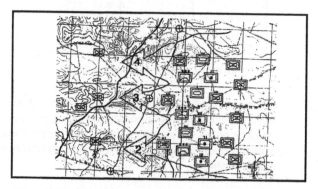

Figure B-III-9. Friendly profile and course of action integration.

(a) Using the C-E constraints and procedures identified from the information provided by the commander, together with data obtained from previous tasks, determine key indicators of friendly operations. For each COA, extract those preparations, activities, or operations that could tip off the threat to the particular COA.

(b) Using a format shown in Figure B-III-8, list the indicators associated with a COA. Any special factors such as operational constraints, optimum weather conditions, or terrain requirements associated with an indicator should be described accordingly.

B-III-8. Review and Validate Threat Assessment Data.

a. Threat assessment data are further refined in order to proceed with the remainder of the vulnerability assessment. The analyst organizes threat data in a format comparable to the friendly forces data. Missing data is identified and requested. The C-SIGINT analyst performs the review and validation of threat data with considerable exchanges of information with other analysts.

b. The procedures for reviewing and validating threat assessment data follow:

(1) Summarize and reorganize threat assessment data.

(a) Compile recent threat assessment information.

(b) Identify information shortfalls.

(c) Coordinate with the collection management section to initiate requests for information.

(2) Extract relevant data for vulnerability assessment.

(a) Extract areas of threat operations most critical to the supported command.

(b) Document threat capabilities and intentions.

(c) Store data for later application.

B-III-9. Identify Vulnerabilities.

a. The analyst compares the intelligence collection threat with the friendly unit susceptibilities to determine the vulnerabilities. Once the vulnerabilities have been identified, the analyst can rank them.

b. The procedures for identifying vulnerabilities follow:

(1) Compare current threat to friendly C-E susceptibilities.

(a) Review indicators of friendly COA.

(b) Use the products developed earlier in the C-SIGINT process to determine where threat capabilities and intentions are directed against susceptible friendly operations.

(c) Determine the probability of threat activity against a friendly C-E operation. Use various statistical and analytical tools. (See the reference list in Technical Bulletin 380-6-1-4.)

(2) Determine which susceptibilities are vulnerabilities.

(a) Designate as vulnerabilities those C-E susceptibilities which are targetable by a specific threat collector.

(b) List (and maintain separately) nontargetable indicators.

(c) Match indicators with threat systems and document specific event characteristics if known; for example, time and location of vulnerabilities.

B-III-10. Rank Vulnerabilities.

a.   The C-SIGINT analyst ranks the vulnerabilities by analyzing them in view of the indicators of friendly COA and EEFI. The ranking is based on criteria estimating the uniqueness, degree of susceptibility, and importance of the vulnerability. The analyst designates the vulnerability as either critical, significant, or important to the success of the overall operation.

b.   The procedures for ranking vulnerabilities follow:

(1)   Establish criteria for measuring the vulnerability. Develop a means for judging whether each identified vulnerability is critical, significant, or important to the success of the operation. These final ratings are attained by evaluating each vulnerability against criteria which address how critical they are to the success or failure of the operation. Uniqueness, importance, and susceptibility to threat are three criteria which measure vulnerability and criticality, and permit an accurate ranking of them. They are defined as follows:

(a)   Uniqueness—the extent to which a vulnerability can be readily associated with a COA.

(b)   Importance—a measure of how critical the vulnerability is to the success of the operation.

(c)   Susceptibility to threat—a measure of the number and variety of threats placed against the indicator.

(2)   Compare vulnerabilities to criteria:

(a)   Combine the criteria and vulnerabilities in a matrix format shown in Figure B-III-10. For each vulnerability, conduct a review against the established criteria. The analysts have in their possession the commander's objectives, prioritized EEFI, and ranking criteria, and can evaluate the vulnerabilities using these data. Vulnerabilities are first rated according to each of the criteria. The horizontal axis of the matrix lists the criteria of uniqueness, importance, and susceptibility.

227

Vulnerability	EEFI	CRITERIA			Overall Numerical Rating
		Uniqueness	Importance	Susceptibility	
Multichannel at DTOC vulnerable to intercept and DF	4(a)2	5	5	4	14
Multichannel at DTOC vulnerable to jamming	4(a)4	3	2	3	8

CRITERIA RATINGS VALUES	OVERALL NUMERICAL RATING
0–2 = Low 3 = Medium 4–5 = High	0—4 = Unimportant 5—8 = Important 9–11 = Significant 12–15 = Critical

Figure B-III-10. Vulnerability matrix format.

(b) List the vulnerabilities on the vertical axis. The degree of satisfaction of a criterion is expressed numerically on a scale of 0 to 5 with 5 being the highest rating. If a vulnerability is highly unique, that is, pertaining to very specialized and infrequently exhibited indicators, it would be assigned a high rating. If the vulnerability is such that it is exhibited in many COA, in many operations, its uniqueness rating would be low (0 to 2).

(1) If a vulnerability is highly important, that is, involving disclosure of a critical EEFI, its rating would be high. An EEFI lower on the commander's list of priorities would receive a lower rating. If the vulnerability is highly susceptible, that is, targeted by numerous threat systems of several types, its rating for susceptibility would be high.

(2) If a single threat system of limited capability is targeting the vulnerability, the rating would be low. The overall ratings are determined by adding the values of the three criteria and placing it under the overall number rating.

(3) Develop ranking.

(a) Once an overall rating is established for each vulnerability, develop a prioritized ranking.

228

Vulnerabilities fall into the broader categories of critical, significant and important, based on the criticality level of criteria satisfied. Vulnerabilities receiving overall ratings between 5 and 8 are considered important; those between 9 and 11 are significant; and those falling between 12 and 15 would be critical.

(b) Enter the list of ranked vulnerabilities in the database. It is retained in hard copy for dissemination, and applied in the countermeasures options development in step three of the C-SIGINT process.

B-III-11. Produce Output From Vulnerability Assessment.

The MDCI analyst presents the vulnerability assessment format shown in Figure B-III-11 as a briefing or a report.

Friendly Supported Unit: 5th Inf Div

1. Situations:

  a. Friendly:

    (1) Mission: Defend to PL Gray, counterattack at 300001z Oct 96.

    (2) Profile statement: See Annex B.

    (3) Indicators of COAs: Multichannel at DTOC.

    (4) Essential elements of friendly information:

      (a) Identification and location of DTOC.

      (b) Loss of C^3 via multichannel.

  b. Enemy

    (1) Intentions: See Annex A.

    (2) Disposition: See Annex S.

    (3) Capabilities: See Annex B.

    (4) Probability of intercept: 90 percent.

Figure B-III-11. Vulnerability assessment format.

2. Prioritized vulnerabilities requiring protection:

   a. Identification of Division Headquarters Element (critical).

   b. Loss of C^3 via multichannel due to jamming (important).

3. Vulnerabilities unable to protect: Loss of C^3 via multichannel due to jamming.

Figure B-III-11. Vulnerability assessment format (continued).

Section IV
Countermeasures Options Development
to
Appendix B
Counter-Signals Intelligence Techniques
and Procedures

B-IV-1. General.

Thus far, our analysis has covered the adversary and our own vulnerabilities. Now it's time to look at our counter-measure options. We need to examine how we can counter the threat's efforts, analyze the risks involved, and present our findings to the commander. Section IV, the third step in the C-SIGINT process, reviews C-E vulnerabilities and identifies, analyzes, prioritizes, and recommends specific options for controlling, eliminating, or exploiting the vulnerabilities.

a.   Countermeasures are required to prevent the exploita-tion of friendly force vulnerabilities by threat systems. The MDCI analyst collects the data and analyzes it to determine possible countermeasures. Many sources are available to the analyst to determine the characteristics of the countermeasures required to achieve the com-mander's objective.

b.   Countermeasures options require the completion of the six tasks shown in Figure B-IV-1.

B-IV-2. Identify Countermeasures Options.

a.   This task is designed to overcome or limit vulnerabilities to the assessed threat.

b.   The procedures for identifying countermeasures options follow:

(1) Collect data.

    (a) Review the vulnerability assessment and list the identified vulnerabilities on the countermeasures options worksheet shown in Figure B-IV-2.

    (b) Extract data on previously used countermeasures for the vulnerabilities and enter the countermeasures options on the countermeasures options worksheet.

    (c) Review current situation for data that would further identify countermeasures options. For example, one commander will enforce strict emission controls to suppress electromagnetic signatures, while another may require continuous and extensive communication.

    (d) List these data on the countermeasures options worksheet.

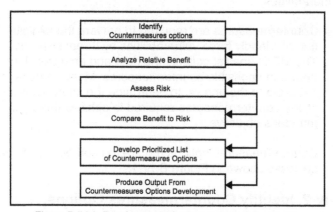

Figure B-IV-1. Development of countermeasures options.

VULNERABILITIES:	COUNTERMEASURES OPTIONS:
Multichannel vulnerable to intercept and DF	Remote equipment
Multichannel vulnerable to jamming	Use destruction
	Place equipment at another echelon
	Use other equipment
	Use deception

Figure B-IV-2. Countermeasure option worksheet.

(2) Identify countermeasures options for each vulnerability.
Use the countermeasures option worksheet.

(a) Prepare a vulnerability to countermeasures matrix
shown in Figure B-IV-3.

(b) List the identified vulnerabilities in the vertical
column and the countermeasures options in the
horizontal column.

Countermeasures options / Vulnerability	Remote equipment	Use destruction	Place equipment at another echelon	Use other equipment	Use deception
Multichannel vulnerable to intercept	X	X	X		X
Multichannel vulnerable to jamming		X		X	X

Figure B-IV-3. Vulnerability to countermeasures matrix.

(c) Match a vulnerability to a countermeasure. This
match is determined by using the identified data
sources.

(d) Check the block that identifies the appropriate countermeasures to the vulnerability.

c. This matrix provides the analyst the countermeasures to
be used for a particular vulnerability.

B-IV-3. Analyze Relative Benefit.

a. This analysis provides the resource requirements for
each countermeasure. The MDCI analyst, in coordination with other staff elements, performs this task.

b. The procedures for analyzing relative benefits of a countermeasure follow:

(1) Identify preferred implementation of the countermeasure. From the identified data sources,
collect data on the preferred countermeasure implementation procedures for each of the

233

countermeasures. Identify the tasks associated with the countermeasure, and gather information about the operational requirements. The following questions will help in gathering this data:

(a) What are the proper start-up procedures for the countermeasure?

(b) What software is associated with the countermeasure?

(c) What steps are involved in operating the countermeasure?

(d) What are the terrain requirements for the countermeasure?

(e) What support services are required?

Countermeasures options / Vulnerability	Remote equipment	Use destruction	Place equipment at another echelon	Use other equipment	Use deception
Multichannel vulnerable to intercept	X	X	X		X
Multichannel vulnerable to jamming		X		X	X

Figure B-IV-3. Vulnerability to countermeasures matrix.

(2) Identify resource requirements. In determining relative benefit, collect data on the resource requirements and command availability of the countermeasure. For example, hardware or personnel required for implementation of the countermeasure is gleaned from the TOE, operator manuals, technical manuals, and other analyst's experience. Additionally, gather information of past experience documented in the CI database and the countermeasures database. The following questions will help in gathering this information:

(a) How many specialists are required?

(b) How many support personnel are required?

(c) What MOS is required?

(d) What are the hardware configurations?

(e) Does the countermeasure require specialized training?

(3) Develop relative benefit table. Upon completion of the data gathering process, enter information on the relative benefit table shown in Figure B-IV-4.

Vulnerability: Multichannel Vulnerable to Interception				
Countermeasures	Resources	Expected Results	Impact on Operations	Shortfalls
Remote multi-channel	Personnel Fuel Vehicles Time Wire for remote Remote equipment	Enemy will think the division is something other than a division	Will take time to set up the remote site Wire has to be guarded for security reasons	Wire comm needs high maintenance May have to replace wire Wire needs to be guarded
Use destruction	Personnel Fire support Ammunition	Complete destruction of threat SIGINT and DF sites	None	Rounds may not be on target SIGINT and DF sites may be moved
Place multichannel at another echelon	Personnel Time Vehicles Fuel Food	Make the threat think the other echelon is the division	SIGINT and DF may not think the division did move	Fired upon before reaching new location New location may not have trained personnel for multichannel equipment
Use deception	Vehicles Deception equipment Trained personnel Time Fuel	Make the threat think we are doing something we are not	Takes a lot of time to plan and implement	Failure to coordinate Equipment may not be available Plan may not work

Figure B-IV-4. Relative benefit table.

(4) Evaluate shortfalls. Evaluate the shortfalls of each of the countermeasures listed and identify alternatives. In the development of shortfalls and alternatives, evaluate the following:

(a) Is the threat vulnerable?

(b) Will the countermeasure reduce or eliminate the vulnerability?

(c) Is deception an effective countermeasure?

(d) Is the countermeasure being developed for training or future use?

(e) Does the countermeasure complement other OPSEC measures?

B-IV-4. Assess Risk.

a. Risk assessment can predict the element of risk to operations when countermeasures are not applied or do not successfully protect friendly vulnerabilities from the threat.

b. The MDCI analyst develops the risk assessment matrix shown in Figure B-IV-5. The procedures for developing a risk assessment matrix follow:

(1) Place a value on the vulnerability and past success of the countermeasure as they apply to specific EEFI. To determine the values, make a judgment based on available information. Use the following scale:

VULNERABILITY	PAST SUCCESS OF COUNTERMEASURE
5 = CRITICAL	5 = HIGH
3 = SIGNIFICANT	3 = MEDIUM
1 = IMPORTANT	1 = MARGINAL
0 = UNIMPORTANT	0 = FAILURE

(2) Fill out the blocks on the matrix as follows:

(a) Block 1: List countermeasures options from countermeasures option list.

(b) Block 2: List specific EEFI.

(c) Block 3: Place a value on the vulnerability of EEFI (5,2,1,0).

(d) Block 4: Place a value on the past success of the countermeasure.

(e) Block 5: Place the numerical risk factor, the following algorithm should be applied: VULNERABILITY - PAST SUCCESS = RISK FACTOR.

(f) Block 6: Annotate the element of risk in Block 6; determine the element of risk by applying the risk factor in Block 5 to the following scale:

| 4-5 = High Risk |
| 2-3 = Medium Risk |

B-IV-5. Compare Benefit to Risk.

a. Having completed the assessment of the risk associated with each countermeasure, the MDCI analyst compares the benefit to the risk for each countermeasure.

b. The procedures for comparing the benefit to the risk follow:

(1) Evaluate benefit.

(a) Using Figure B-IV-4, compare the expected result from the countermeasure implementation with its impact on operations and resource requirements. For example, if the expected results for implementation of the countermeasure are considered high, the impact on operation low, and few resources are required, the expected relative benefit will be high.

(b) Conversely, if the expected result is low, the impact on operations is high, and the counter-measure requires large resources, the relative benefit should be considered low. A value of Low, Medium, or High is then placed on the benefit to risk form shown in Figure B-IV-6.

COUNTERMEASURES OPTION	BENEFIT	RISK
Remote equipment	Medium	Low
Use destruction	High	Low
Place equipment at another echelon	Low	Medium
Use deception	Medium	High

Figure B-IV-6. Benefit to risk form.

(2) Evaluate risk. Review Figure B-IV-5. Extract the risk assessment from Block 5, and enter the value in Figure B-IV-6. This completed form provides the risk associated with the relative benefit of each countermeasure.

B-IV-6. Develop Prioritized List of Countermeasures Options.

a. This list provides the commander and staff with recommended countermeasures options for identified vulnerabilities.

b. The procedures for developing a list of prioritized countermeasures options follow:

    (1) Prepare countermeasures effectiveness.

        (a) Using the Benefit column (Figure B-IV-6), list all countermeasures options in order from the most to the least effective on the countermeasures effectiveness to costliness worksheet shown in Figure B-IV-7. For example, if destruction, remoting, deception, or moving the equipment to another echelon were your options, your effectiveness column would look like the following:

            i.   Use destruction.

            ii. Remote equipment.

            iii. Use deception.

            iv. Place equipment at another echelon.

(b) The number in front of the countermeasure is now the effectiveness value of that countermeasure.

(2) Prepare countermeasures costliness.

(a) Using the Risk column in Figure B-IV-6, list all countermeasures options in order from the least to the most costly in the Costliness column (Figure B-IV-7). For example, using the same countermeasures options used in b(1) above, your costliness column would look like the following:

   i.   Remote equipment.

   ii.   Use destruction.

   iii. Place equipment at another echelon.

   iv.   Use deception.

EFFECTIVENESS	COSTLINESS	CM RATING AND PRIORITY
1. Use destruction.	1. Remote equipment.	1. Use destruction.
2. Remote equipment.	2. Use destruction.	2. Remote equipment.
3. Use deception.	3. Place equipment at another echelon.	3. Use deception.
4. Place equipment at another echelon.	4. Use deception.	4. Place equipment at another echelon.

Figure B-IV-7. Effectiveness to costliness worksheet.

(b) The number in front of the countermeasure is now the costliness value of that countermeasure.

(3) Prepare countermeasure priority list.

(a) Using Figure B-IV-7, add the number from the Effectiveness column to the number in the Costliness column. Determine the

239

countermeasures which are the most effective
and the least costly, and produce a prioritized
list of countermeasures. The lower the rating
the higher the priority the countermeasure has.
When there is a tie, the countermeasure that
has the higher effectiveness rating is given the
higher priority. For example, using the information in paragraph B-IV-6(1) and (2), the
countermeasure rating column would look like
the following:

i.   Use destruction. 3.

ii, Remote equipment. 5.

iii. Use deception. 5.

iv. Change echelon. 7.

(b) Using this example, the countermeasure
prioritized list looks like that in Figure B-IV-7.

B-IV-7. Produce Output From Countermeasure Options Development.

The countermeasures option process is complete when
the analyst reviews and recommends the countermeasures
options to the operations element.

Section V
Countermeasures Evaluation
to
Appendix B
Counter-Signals Intelligence Techniques and Procedures

B-V-1. General.
Commanders determine which countermeasures are to be applied. Once applied, it is the MDCI analyst's job to evaluate the countermeasure's effect. Section V, the last step in the C-SIGINT process, determines how well the applied countermeasures worked and their impact on the operation.

a. Lessons learned provide feedback to the commander and serve as information for other commands considering similar countermeasures options. Countermeasures evaluation is the review, analysis, and evaluation of countermeasures to determine their effectiveness. The evaluation includes five major types:

    (1) C-SIGINT database.

    (2) Intelligence data.

    (3) Interviews.

    (4) Reviews of messages, reports, and other operational documentation.

    (5) Reviews of actual profiles during the operation.

b. The specific tasks in the countermeasures evaluation format are shown in Figure B-V-1.

B-V-2. Validate Commander's Guidance.

a. Since countermeasures are planned in accordance with the commander's guidance, operational deviations from the commander's guidance may affect their effectiveness, even though the countermeasures are performed as planned. The first task validates the commander's guidance. To ensure the proper baseline is applied in evaluating the countermeasure, the MDCI analyst reviews the commander's guidance for changes or misunderstandings.

b. The procedures for validating the commander's guidance follow:

(1) Review the commander's guidance and EEFI. Retrieve the commander's guidance and objectives collected and stored during the vulnerability assessment. The EEFI statement and friendly COAs developed during the vulnerability assessment are also important sources. Review the OPLAN, OPORD, and EEFI; add information or reports unavailable during or produced after the vulnerability assessment; and update the statement of the commander's operational concept generated during the vulnerability assessment.

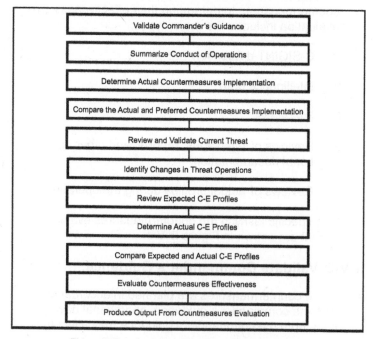

Figure B-V-1. Countermeasures evaluation format.

(2) Verify guidance and EEFI. Present the updated summary of the commander's operational concept to the operations staff. They review the summary, and note any misinterpretations or information gaps. The analyst reviews the operations staff's comments and completes the final verified statement of the commander's guidance and EEFI.

B-V-3. Summarize Conduct of Operations.

a. During this task the analyst compares how well actual operations matched planned operations. If the countermeasure is specific to an operation, the summary occurs upon completion of the operation. If the countermeasure is part of normal peacetime operations, the summary occurs at regular intervals. The MDCI analyst directs the evaluation. The analyst coordinates with the appropriate staff elements and talks to participants in the operation.

b. The procedures for summarizing the conduct of operations follow:

(1) Determine major activities. Use the commander's operational concept identified in paragraph B-V-2b(2) and the commander's guidance to identify the major activities required to conduct operations. Seek answers to the following questions:

(a) What type of task is it: (C 2 support)?

(b) Who was responsible (officer, NCO)?

(c) Who performed each of the tasks?

(d) What equipment was used in implementing the task in supporting the operation?

(e) What other units were involved?

(f) What supply channels were used?

    (g) Was the task part of normal or special operations?

    (h) Was the situation hostile or peaceful?

(2) Identify sources and collect data:

    (a) Review the data requirements and identify the probable sources of data. Most frequently, the first source of information is written reports, memorandums, or journals of an operation.

    (b) Develop a data collection format using the characteristics as key for describing the operation shown in Figure B-V-2.

    (c) Identify personnel or units best suited to gather the data, and request they be tasked to collect data.

    (d) Use organic resources to collect data not coming from tasked sources.

    (e) List shortfalls in meeting the data requirements as gaps after completing the data collection.

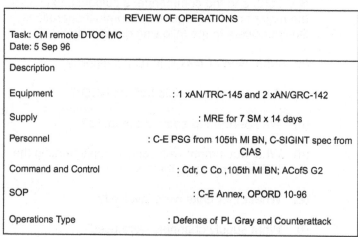

REVIEW OF OPERATIONS	
Task: CM remote DTOC MC Date: 5 Sep 96	
Description	
Equipment	: 1 xAN/TRC-145 and 2 xAN/GRC-142
Supply	: MRE for 7 SM x 14 days
Personnel	: C-E PSG from 105th MI BN, C-SIGINT spec from CIAS
Command and Control	: Cdr, C Co ,105th MI BN; ACofS G2
SOP	: C-E Annex, OPORD 10-96
Operations Type	: Defense of PL Gray and Counterattack

Figure B-V-2. Data collection format.

(f) Interview personnel who took part in the operations. Usually, your information needs are met by analyzing operations.

(g) May request permission to interview others identified as key personnel in the reports. Enter the results of the interview on the form.

(3) Review and finalize data:

(a) Review the data, assuring all requirements are met.

(b) Request additional information from the operations officer if conflicting information exists.

(c) Enter final information on the form.

(4) Compare and summarize data about conduct of operations:

(a) Complete a final review of the data collected.

(b) Compare the actual conduct of operations with the OPLAN and OPORD.

(c) Note where operations proceeded as planned and where they deviated from plans.

(d) Write a summary statement of the differences for later analysis.

B-V-4. Determine Actual Countermeasures Implementation.

a. The MDCI analyst gathers information about countermeasures implementation operations. The analyst can collect information about implementation any time after the countermeasure has been initiated. The preferred scheme is to collect information about implementation at regular intervals.

b. If the countermeasure is part of general peacetime operations, evaluations of the countermeasure operations should be ongoing. If the countermeasure is specific to an operation, the minimum number of evaluations is two. Accurate information about countermeasures implementation is necessary for evaluating how well the countermeasures were implemented. The analyst may request information from the operations staff, adjacent or higher echelon assets supporting a command, and participants in an operation.

c. The procedures for determining actual countermeasures implementation follow:

(1) Determine data requirements to implement countermeasures:

(a) Determine the preferred countermeasures implementation procedures.

(b) Examine the preferred tasks and expected results, responsibilities, equipment, operational characteristics, weather and terrain impacts, and effects of the countermeasure on adjacent or higher echelon commands.

(c) Enter the information on the countermeasures implementation worksheet shown in Figure B-V-3. Specific questions must be tailored to each situation.

(d) Answer the following types of questions to determine data requirements:

(1) Are there SOPs for the operations? Are they affected by other countermeasure operations?

(2) Who are the personnel responsible for the countermeasure? Have they performed the requisite duties?

        (3) What are the specific tasks? What opera-
tions will be affected? By whom? In what
time frame?

        (4) What are the expected results? Will they
affect operations?

(2) Identify data sources. Review the data requirements and determine what data are stored in the
database; what new data are required; and the best,
most complete, and valid sources for the new data.
Written reports, memorandums, or findings of an
operation are a primary source information. The
analyst must verify and supplement this documentation from—

    (a) The analyst's own experience.

    (b) Interviews of participants in the operation.

    (c) Interviews of staff members.

    (d) Reviews of the actual conduct of operations.

    (e) Queries of other commands in the theater.

(3) Collect and summarize data:

    (a) Review the C-SIGINT database and other docu-
mentation to fulfill the data requirements.

    (b) Identify data supporting the needs and enter the
data in the Findings column of Figure B-V-3.

    (c) Enter the data sources in the last column. Upon
completion of this first data survey, identify
information shortfalls by gaps in the rows.

    (d) May have to request information from other
sources, such as the operations staff, adjacent
commands, or participants in the operation.

    (e) Review for completeness before beginning the next
task.

B-V-5. Compare the Actual and Preferred Countermeasures Implementation.

a. The comparison of actual and preferred implementation shows successes and shortfalls in countermeasures actions. The MDCI analyst uses these successes and shortfalls in later evaluations of countermeasures options to manage ongoing countermeasures activities and to determine whether new vulnerabilities exist.

b. The comparison may be performed during an ongoing operation or upon completion. The completion provides needed information and analysis about countermeasures implementation in different operational situations. The comparison is based on information generated during the identification of countermeasures options and the internal processes in countermeasures evaluation.

c. The procedures for comparing the actual and preferred countermeasures implementation follow:

(1) Review data on countermeasures application. Review the data collected and organized in the development of countermeasures options and in the previous implementation. Simple countermeasures will have limited support and few activities or tasks; more complex countermeasures, such as deception, have numerous activities and relationships. In the latter case, analyze key activities first. If they are not met, the countermeasures are not effective and subsequent analysis is unnecessary. For example, the remoting of the multichannel supporting the Division Tactical Operation Center (DTOC) requires a separate site with NCO supervision. The personnel components of the evaluation are, therefore, examined first, since they represent the largest impact on the determination of effectiveness.

(2) Estimate the degree preferred countermeasures actions were completed. Determine the degree of completion by comparing the actual with preferred countermeasures performance.

(a) Match actual and preferred:

    i.   Review the preferred countermeasures implementation from the analysis of the relative benefit.

    ii.   Match data from the actual implementation with the preferred implementation.

    iii.   Review the countermeasures implementation review worksheet shown in Figure B-V-4, focusing first on the key activities.

    iv.   Identify areas where the actual countermeasures activities were similar to the preferred actions.

CM : Remote DTOC MC Date: 4 Sep 96			
TASKS		**PERCENT COMPLETE**	**REMARKS**
Key :	Allocate MSE to provide communications with remoted MC.	100	MSE provided by C Company, 105th MI.
Median :	Provide NCO supervision at remote site.	50	DTOC and corps MCs had NCOs, but the MSE were operated by PFCs. MSE operators were not efficient in sending msgs.
Support :	C-E Plt Sgt reviewed division logs, but not corps logs due to time.	70	Additional data gatherer needed.
	MREs were consumed twice as fast as expected due to additional corps personnel.	50	Conduct additional liaison with corps and adjacent divisions before next implementation of this CM to ensure sufficient MREs are available.

Figure B-V-4. Countermeasure implementation review worksheet.

    v.   Estimate the percent of the preferred completed actions.

    vi.   If the actual countermeasure is at least 50 percent of the preferred countermeasures operations, enter the countermeasure in the implementation review worksheet.

vii. If the countermeasures have not been implemented as planned, according to the preferred operations, the percent completion of the standard is determined to be zero, and the analysis is complete.

(b) Develop an overall estimate of completion of the preferred countermeasures operations. For multitask countermeasures or situations including more than one countermeasure—

i. Review the individual key estimates listed in Figure B-V-4.

ii. Review support activities to determine the general level of success and failures.

iii. Look for trends and specific cases decreasing full countermeasures activity.

iv. Find the median percent estimate and describe the level of overall success using the median and a verbal description of general support.

B-V-6. Review and Validate Current Threat.

a. Validation of the current threat provides a baseline for subsequent analysis of changes in threat operations. For short-term countermeasures operations, the validation should be completed before evaluation of the countermeasures implementation. The analyst must work with the best and most current threat data available. Validated threat is a necessary baseline for evaluation of threat SIGINT/EW operations, and, ultimately, the effectiveness of countermeasures.

b. The operations staff receives and analyzes current information to generate the threat assessment. The information is requested by and transferred to the MDCI analyst for countermeasures evaluation. The procedures for reviewing and validating current threat follow:

(1) Review expected impacts on threat.

    (a) Review the objectives and expected results of the countermeasure. Identify any expected impacts of the countermeasure on the threat by asking the following types of questions:

        i. Is the countermeasure objective to impair or destroy the capabilities of a known collector rather than to protect friendly EEFI?

        ii. Is the threat able to redirect assets toward another friendly vulnerability?

        iii. Have similar countermeasures worked in other situations against similar threats?

        iv. How important is the collector to the operations or doctrine of the threat?

        v. Is the threat asset an EEFI for friendly operations?

    (b) Then complete the review of information by identifying specific threat assets mentioned as targets, and listing these targets for further analysis.

(2) Review threat assessment. Review the threat assessment to determine whether the threat targets are examined and evaluated in the product. If included, quickly review supplemental intelligence reports. If not included, request additional intelligence support to complete a threat review specific to the targets.

(3) Review other intelligence products about threat actions. As necessary, identify recent intelligence reports describing changes in location, capabilities, and intentions of the threat systems.

(4) Validate the threat assessment:

(a) Compare the current threat assessment with the previous threat assessment.

(b) Prepare a list of the existing pre-countermeasures implementation threat.

(c) Validate the list of key indicators by reviewing them for consistency, checking the sources, and if necessary, requesting assistance from the operations staff.

B-V-7. Identify Changes in Threat Operations.

a. Current intelligence data collected after countermeasures implementation are used to estimate the effects of implemented countermeasures on the targets identified shown in Figure B-V-5. The MDCI analyst performs this process after the countermeasure has been applied, if it is a shortterm operation. In a continuing operation, the analyst performs the process at regular intervals to identify significant changes potentially affecting other countermeasures or operations. If the countermeasure is directed against the threat, the analyst must determine whether the threat has changed and whether friendly vulnerabilities have changed. The analyst integrates the threat information into the countermeasures evaluation.

	PRE-CM THREAT INDICATORS	POST-CM THREAT INDICATORS	REMARKS	PERCENT CONFIDENCE LEVEL
			CM: Remote DTOC MC Date: 10 Nov 96	
Capabilities	Intercept and DF of MC	Increased IMINT and HUMINT		80
Operations	Low level of threat SIGINT Low SIGINT site movement	Increased traffic in threat C^3 Five site movements in a week		80
Intentions	Attack imminent MC jammer in proximity to MBA	No jamming reported for 8 days		100

Figure B-V-5. Countermeasures impact on threat worksheet.

b. The procedures for identifying changes to the threat operations follow:

(1) Compare current and assessed actions. Compare the most recent information in situation reports, intelligence reports, messages, or verbal conversations with the validated threat assessment.

(a) Review the list of threat indicators identified in Figure B-V-5 and the expected impact of the countermeasure on the threat:

i. Examine current intelligence information about the indicators.

ii. Examine capabilities, such as the organization of threat resources, their readiness, deployment, equipment, and the introduction of new capabilities to the opposing unit.

(b) Enter any changes in Figure B-V-5. Include any remarks about the implications of the changes, including whether the changes correlate with the expected result of the countermeasure. For example, you receive an intelligence report describing a decrease in threat attempts to intercept allied tactical communications because of electronic protection (EP) built into new radios. The implication is that the friendly vulnerabilities have decreased, causing a change in threat capabilities. The level ranges from 50 percent for a guess to 100 percent for certainty. Higher confidence levels are given for more specific information about an indicator, multiple sources of information, multiple disciplines, or corroboration by another analyst.

(c) Next, review the current intelligence to determine whether any information suggests change to other operations. Examples might include increased IMINT, more active HUMINT, sabotage, or other active pursuits. The redirection of threat efforts increases the estimate and confidence in countermeasures effectiveness against their target. (Although the redirection may also mean changed friendly vulnerabilities.) Update Figure B-V-5.

(d) Review and evaluate intentions:

 i. Review the threat assessment for past intentions and examine the current intelligence for indicators of threat intentions.

 ii. Identify changes by examining whether any current intelligence documents changed intentions and by examining patterns of behavior to see whether they match doctrine. You can attribute the changes in operations to the countermeasures if the changes cannot be explained by doctrine or no documentation of changes exists.

 iii. Include a confidence level for the interpretation. For example, personnel redeployment following the implementation of countermeasures.

 iv. Request assistance, if needed, from other members of the intelligence or operations staff knowledgeable about doctrine and plans to help identify changed intentions.

(e) After you complete the identification of changes in the threat, review the results listed in Figure B-V-5. All areas with consistent information about threat changes are highlighted.

Any contradictions are noted. Attempt to resolve any contradictions by reviewing data or by requesting assistance from intelligence sources.

(2) Identify and evaluate the significance of changes. A second analysis follows the determination of any changes. Determine how significant the changes are: whether they represent a major change of threat capabilities and intentions, and the overall importance of the change.

   (a) Compare past examples of changes in operations, capabilities, and estimate of intentions with the current data. The comparison uses current information compiled in Figure B-V-5, past compilations, and supplementary documentation stored by the analyst. Then review the documents, collect similar threat indicators (for example, types of collectors), and list indicators for threat changes by countermeasures review worksheet shown in Figure B-V-6. Complete this form by entering descriptions of findings. In the search for changes in patterns, ask specific questions about particular threat systems and countermeasures:

      i.   Does the threat tend to alter collection operations consistently (for example, redirecting assets toward another target or using different assets toward the supported commander's EEFI)?

      ii. Are threat changes matched with the type of friendly situation?

      iii. Do threat changes occur after a certain time period of countermeasures application or are they immediate?

      iv. Does the threat attempt to use EA or EP?

(CLASSIFIED WHEN FILLED IN)					
	Current Threat Changes		Past Changes		
Date	Description	Situation	Description	Situation	Source

Figure B-V-6. Format for threat changes by countermeasures review worksheet.

   (b) Use the results from the evaluation of the countermeasures impacts on the threat in the final evaluation of threat indicators. Estimate the significance of a change in threat capabilities and intentions by examining the following five characteristics:

      i. Scope of change (how wide is the impact)?

      ii. Effects of change (how many assets does the change affect)?

      iii. Level of change (at what echelon in the chain-of-command)?

      iv. Direction of type of change (what COA was taken)?

      v. Overall evaluation of the significance of the change (in comparison with historical data filed in CI database).

(c) Enter each change documented in Figure B-V-5 in the significance of threat changes worksheet shown in Figure B-V-7. Then determine the magnitude of each change (based on a comparison of the previous actions and the actual change identified in the threat indicators).

Classified when filled in							Date: _____	
	A	B	C	D	E	F	G	H
Capabilities								
Operations								
Intentions								

Figure B-V-7. Format for significance of threat changes worksheet.

(d) Use a scale of 1 to 5 for ranking the change of each of the six indicators. If the change is large, enter a 5; if small, enter a 1. Then compile and review the individual assessments; a significant change is one in the 4 to 5 range in at least three of the five characteristics or any composite change factor totaling at least 15.

(e) The final step is to estimate the confidence of the assessment of change. Use the 50 to 100 percent scale using the following guide to determine the confidence level:

    i. Multiple sources used.

    ii. Multiple disciplines applied.

    iii. Multiple analysts agreeing with the assessment.

    iv. An assessment of the changes in threat operations due to implementation of the countermeasures.

(f)   Confidence is high if two of the above criteria are met. Confidence is low if none are met. The confidence level is then entered in Figure B-V-7.

(3)   Summarize the analysis in a two-part format. The first part is a short summary of the applied countermeasures, current intelligence reports, and an assessment of the areas of threat change due to countermeasures implementation. The second part includes the detailed estimate of changes presented in Figures B-V-5 and B-V-7.

B-V-8. Review Expected C-E Profiles.

a.   The analyst requires a base line of expected friendly profiles to determine whether a countermeasure has had the expected effect on the profiles by reviewing expected patterns and signatures. The data are initially collected to perform the vulnerability assessment and are applied during this process.

b.   The MDCI analyst—

(1)   Reviews expected patterns and signatures by identifying profile information associated with the countermeasures during the validation of the commander's guidance and the summarization of the operation.

(2)   Searches the friendly forces database for information about the equipment and operational profiles associated with the countermeasures.

(3)   Enters countermeasures-related profile data collected into the profile comparison worksheet shown in Figure B-V-8.

(4)   Checks to ensure data are current and complete. If not, requests additional information from the operations and logistic sections.

	EXPECTED		ACTUAL	
			CM: Remote MC Date: 10 Nov 96	
	Command and Control	Operations	Command and Control	Operations
Physical Signature	2 x M35A2	1 x M104 1 x M35A2	1 x M35A2 1 x M1008	1 x M104 1 x M35A2
Electronic Signature	AN/VRC-46 AN/TRC-145	AN/GRC-142 AN/MRC-108B AN/GRC-106	AN/VRC-46 AN/GRC-142	AN/GRC-142 AN/MRC-108B AN/GRC-106
Pattern Data	MC to DISCOM and DIVARTY		Additional MSE for communication with remote MC	

Figure B-V-8. Profile comparison worksheet.

B-V-9. Determine Actual C-E Profiles.

a. The analyst requires valid data about actual C-E profiles after countermeasures application to evaluate the successful change of a command's C-E profile. The analyst needs specific data for comparison with expected profiles to evaluate countermeasures. The analyst updates and maintains the friendly forces C-E profile information. Data sources include the CI database, interviews with personnel, reviews of reports, and review of operations.

b. The procedures for determining actual C-E profiles follow:

   (1) Determine data requirements.

      (a) Identify specific data requirements to compare the actual and expected profiles. Review this data and ask the following types of questions:

         i.   What types of equipment are used?

         ii. What logistic channels are used?

         iii. Are there SOPs?

         iv. What doctrinal requirements result in profiles?

259

      (b) Identify specific questions for each counter-
measures evaluation.

  (2) Collect data about actual C-E profiles. Identify
command-specific sources of data to fill the requirements. Common sources include actual participation in operations, new evaluations of equipment
and procedures resulting from site survey or new
equipment information, interviews of personnel, and
reviews of operational reports. Begin by—

      (a) Reviewing reports.

      (b) Identifying information about actual profiles.

      (c) Matching the data requirements.

      (d) Entering data in Figure B-V-8.

  (3) Summarize data. After preliminary reviews, information gaps remaining are filled by other sources.
This depends on the applied countermeasures and
resources available (sufficient personnel and equipment vital to the determination of C-E profile are not
available and require EAC augmentation). Verify the
descriptions of the profile by using data from at least
two sources.

B-V-10. Compare Expected and Actual C-E Profiles.

a.   The analyst compares what should happen with what
does happen to determine whether the countermeasures were effective and whether stated countermeasure
results are valid. The comparison contrasts expected
C-E profiles, based on historical data, with actual
profiles. The analyst performs the comparison and
analysis using the CI database and the data collected in
paragraph B-V-9.

b.   The procedures for comparing expected and actual C-E
profiles follow:

(1) Analyze actual command profile:

   (a) Review the data generated in paragraphs B-V-8 and B-V-9.

   (b) Compare the expected and actual C-E profiles.

   (c) Summarize the differences for each attribute.

   (d) Note the differences between expected and actual profiles.

   (e) Review the comparison to answer the following questions and include these findings in a report to the commander.

      i. What attributes most closely matched expected profiles?

      ii. Were any elements in an organization successful in matching expected C-E profiles?

      iii. Have adjacent commands or OPSEC policies affected the C-E profiles?

(2) Analyze the relationship between countermeasures expected profile, and actual profile.

   (a) Complete an additional evaluation of the countermeasures and profiles to provide an overall review of the countermeasures effect on C-E profiles and to identify additional vulnerabilities as a result of the countermeasures. Note the continuing C-E profile vulnerabilities; list new vulnerabilities identified in the countermeasures review; and document the evaluation.

   (b) This information becomes part of the recurring lessons-learned reviews conducted at theater level and is stored locally for continued evaluation.

B-V-11. Evaluate Countermeasures Effectiveness.

a. The final process reviews and combines all previous countermeasures evaluations. Both ECB and EAC elements use the information to modify ongoing countermeasures operations, as summary information for subsequent countermeasures options development, and for training analysts.

b. The procedures for evaluating countermeasures effectiveness follow:

    (1) Evaluate the success of countermeasures in meeting specific objectives defined in paragraph B-V-2. Review the findings of each countermeasures evaluation, listing the cases where results of countermeasures implementation met the expectations identified during development of countermeasures options. Compare these successes with the number and type of failures. Repeat this procedure for each type of analysis. A scale for comparing success to failure follows:

        (a) 0 percent - Failure.

        (b) 25 percent - Marginal success.

        (c) 50 percent - Moderate success.

        (d) 75 percent - Substantial success.

        (e) 90 percent - High level of success.

    (2) Evaluate overall success of countermeasures. Using the countermeasures finding worksheet shown in Figure B-V-9, enter the estimates of success for each of the countermeasures evaluation criteria. For each countermeasure, review the level of success per objective. Count the number of cases where the countermeasure has a greater than 50 percent success rate and divide this number by the total number of cases. The resulting ratio provides an overall

estimate of success, supplemented by specific success rates for particular objectives estimated in paragraph B-V-11(1).

Vulnerability: intercept and DF of DTOC MC				Date: 10 Nov 96
CM APPLIED	**PERCENT ESTIMATE OF CM SUCCESS**	**STRENGTHS**	**WEAKNESSES**	**PERCENT TOTAL SUCCCESS RATIO**
Remote MC	50	Causes threat to move often, increasing the chances of friendly observation	Does not remove vulnerability increases amount of MSE traffic	50

Figure B-V-9. Countermeasures finding worksheet.

(a) Complete a listing of strengths and weaknesses of a given countermeasure in the situation examined.

(b) Use the definition of the situation from paragraphs B-V-2 and B-V-3.

(c) Develop and maintain a list of countermeasures strengths and weaknesses to support countermeasures development in subsequent operations.

(4) Identify assumptions for revalidation. Compare current findings with previous countermeasures analyses. This helps identify continuing problems that require review. For example, regular changes of call signs may have been a sufficient countermeasure in a particular location because the threat was unable to intercept clear voice and was unable to identify stations within the net with any degree of accuracy. Changes in technology, threat operations, and net radio traffic may cause the effectiveness rating of the applied countermeasure to decline. The analyst must recommend a review of the countermeasure as threat operations become more successful. Continual revalidation is a necessary procedure.

(5) Summarize findings. Summarize findings for quick responses to the commander; and produce a lessons-learned review for EAC and theater.

B-V-12. Produce Output From Countermeasures Evaluation.

Output may consist of reports, briefings, or messages. Many of the forms used as working aids in CI analysis may be attached as summary statements.

Appendix C

Counter-Imagery Intelligence Techniques and Procedures

C-1. General.

The proliferation of imagery systems worldwide, especially the platforms carrying imagery systems, makes the task of the C-IMINT analyst much more complicated than ever before. Relatively inexpensive platforms that are easily transported and operated, such as unmanned aerial vehicles, are becoming available to anyone who wants to employ them. For the more sophisticated, there are other platforms either continuously circling the planet or in geosynchronous orbit, available for hire by anyone with the desire and the ability to pay the freight. An adversary need not possess the technology to build and launch such a platform. He merely buys time from the operators of the platform and obtains the products acquired during his allotted time. Like all other CI functions, C-IMINT depends on the analyst knowing the adversary and knowing ourselves. It begins long before friendly forces deploy for any operation and continues throughout the operation. It goes on even after our forces return to their home station after completion of the operation.

C-2. Operations.

C-IMINT begins with knowledge. The MDCI analyst must have a thorough knowledge of the threat in the objective area and any threat from outside the AO that may influence our operations.

    a. Predeployment. Prior to any operation, the MDCI analyst needs to prepare indepth. In addition to researching data on the threat and the AO, the MDCI analyst gathers information and builds a database to serve C-IMINT in the coming operation. During this phase, the MDCI analyst initiates quick reference matrices and the IMINT situation overlay.

(1) Adversary intelligence flight matrix. These matrices are concerned with other platforms used by the adversary. Tracking these collection systems continuously allows the analyst to analyze threat IMINT collection patterns.

(2) System component quick reference matrix. These matrices are concerned with adversary system's capabilities and processing times. This file is part of the database which equates to an OB file on threat IMINT systems shown in Figure C-1.

(3) IMINT situation overlays. These are the paths of adversary intelligence collection flights depicted on the friendly operations graphics. They identify areas susceptible to collection.

b. Friendly patterns. Pattern analysis is the detailed study of friendly activities to determine if a unit performs the activities in a predictable manner, thus creating a monitorable pattern of activity. These actions cue an observer to a unit's type, disposition, activity, and capability. Imagery coverage of the AO is essential for planning and for reference later during operations. Small or intermediate scale imagery covering the entire AO may be obtained from general reference files or national sources and need not be newly flown. The presence of US reconnaissance aircraft making numerous passes over territory belonging to another nation would tip off an impending operation. Therefore, file imagery or imagery obtained by satellite may be the only reference available.

SYSTEM COMPONENT QUICK-REFERENCE MATRIX					
SYSTEM:			DATE:		
ORGANIZATION	LOCATION	CHARACTERISTICS	STRENGTH	TACTICS	REMARKS

Figure C-1. System component quick-reference matrix.

(1) Friendly IMINT is used, when available and of high enough priority, to determine friendly patterns which may be susceptible to IMINT collection. These patterns are key indicators to the enemy of specific operational activities. Patterns usually occur because of a unit's SOP and doctrine. Example patterns include—

    (a) Relocating fire support units forward before an attack.

    (b) Locating TOCs in the same relative position to maneuver elements and to each other.

    (c) Repeating reconnaissance overflights of areas planned for ground or air attack about the same time before each operation.

(2) Information gained from imagery provides a means of checking other reports and often produces additional detailed information on a specific AI. All friendly activities thus need to be examined collaterally with imagery of a particular area. Imagery can provide confirmation of installations, lines of communications, and operational zones. SLAR, for example, can detect night movements of watercraft.

(3) Finally, in the overall evaluation, analysts synthesize the separate trends developed during analysis. Such a process identifies the possible compromise of an existing element, activity, or characteristic based on logical relationships and hypotheses developed by analysis. The pattern analysis technique is just one of many techniques designed to help evaluate friendly units for vulnerability to threat IMINT. The process is a continuous one.

(4) Analysis of a unit's movements gives significant clues to its intentions, capabilities, and objectives. By applying this technique against our own units, analysts can identify vulnerabilities. Movement analysis forms an important step in the identification and recommendation of countermeasures.

    (a) SLAR is a primary sensor in detecting moving targets or moving target indicators (MTIs) and is usually associated with the special electronics mission aircraft and Joint STARS platform. While the sensor is primarily focused at enemy MTIs, it can be used to identify friendly movement patterns that may also be collected by the enemy.

(b) The tracks created by a unit can give excellent indication of a unit's disposition. Any time a unit moves away from hard packed roads, the danger of leaving track signatures is very high. There are certain countermeasures which should be observed to disguise or eliminate these signatures:

   i. Conceal tracks by netting or other garnish.
   ii. Disperse turnouts near CPs.
   iii. Place installations and equipment near hard roads where concealment is available.

(c) Our IMINT resources can determine the effectiveness of a friendly unit's program to suppress its visual and thermal signatures, including positioning of assets. Friendly aerial reconnaissance is extremely limited and must be planned for well in advance. The following are examples of countermeasures that could be used to reduce our vulnerability to enemy IMINT:

   i. Using traffic discipline when moving into and out of the installation. This may require walking some distance to a CP.
   ii. Driving in the tree lines when roads are not available.
   iii. Extending new roads beyond the CP to another termination.
   iv. Controlling unauthorized photographic equipment.
   v. Using physical security measures to prevent optical penetration.
   vi. Using proper camouflage procedures.
   vii. Limiting the dissemination of photographs made from within the installation.
   viii. Avoiding the use of direction signals and other devices which provide information.
   ix. Concealing equipment markings.

x. Preventing detection by infrared imaging (nets, infrared generators).

xi. Eliminating open-air storage of special equipment, raw materials, and telltale objects.

(d) The key to proper positioning of assets on the ground is to use natural features as much as possible. Obvious locations such as clearings may be more convenient but should be avoided at all costs. This includes night operations. Infrared and SLAR missions are particularly effective at night. Units should be well dispersed since a high concentration of tents and vehicles, even well hidden, will stand out on imagery to a trained analyst.

c. **Evaluation of countermeasures.** For these countermeasures to be effective, every command should develop a self-evaluation system to ensure proper employment.

Glossary

Section I. Abbreviations and Acronyms

A

ACCO	Army Central Control Office
acct	account
ACE	Analysis and Control Element
ACofS	Assistant Chief of Staff
ACR	armored cavalry regiment
ADA	air defense artillery
ADP	automated data processing
ADPSSEP	automatic data processing system security enhancement program
aer	aerial
AI	area of interest
AKA	also known as
AO	area of operations
AR	Army regulation
ARNG	Army National Guard
ARSOF	Army Special Operations Forces
ASAS	All-Source Analysis System
ASP	ammunition supply point
ASPP	Acquisition Systems Protection Program
attn	attention
AWOL	absent without leave

B

BDA	battle damage assessment
BE	basic encyclopedia
BI	background investigation
bn	battalion
BOS	Battlefield Operating System BSA brigade support area

C

C 2	command and control
C 3	command, control, and communications
CA	Civil Affairs

CARVE	criticality, accessibility, recuperability, vulnerability, and effect cdr commander
C-E	Communications-Electronics
CE	counterespionage
CFSO	CI force protection source operations
CG	commanding general
C-HUMINT	counter-human intelligence
CI	counterintelligence
CIA	Central Intelligence Agency
CIAC	CI Analysis Center
CIAS	counterintelligence analysis section
CIDC	Criminal Investigation Command
C-IMINT	counter-imagery intelligence
CINC	Commander in Chief
CM	countermeasures
cmd	command
co	company
COA	course of action
COMINT	communications intelligence
comm	communications
COMSEC	communications security
CONUS	continental United States
CP	command post
C-RISTA	counterreconnaissance, intelligence, surveillance, and target acquisition
C-SIGINT	counter-signals intelligence
CSP	CI scope polygraph

D

DA	Department of the Army
DACAP	DA Cryptographic Access Program
DCID	Director of Central Intelligence Directive
DCII	Defense Central Index of Investigations
DCSINT	Deputy Chief of Staff, Intelligence
demo	demonstration
DF	direction finding
DHS	Defense HUMINT Services
DI	deception indicated
DIA	Defense Intelligence Agency
DIAM	Defense Intelligence Agency Manual
DIS	Defense Investigative Service

DISCOM	Division Support Command
div	division
DIVARTY	division artillery
DM	Deutsche Mark
DOD	Department of Defense
DODD	Department of Defense Directive
DPA	Data Processing Activity
DPOB	date, place of birth
DS	direct support
DSO	defensive source operations
DTAC	CP division tactical CP
DTG	date-time group
DTOC	Division Tactical Operations Center

E

EA	electronic attack
EAC	echelons above corps
ECB	echelons corps and below
EEFI	essential elements of friendly information
ELINT	electronic intelligence
ELSEC	electronic security
EO	executive order
EOB	electronic order of battle
EP	electronic protection
EPB	electronic preparation of the battlefield
EPW	enemy prisoner of war
ES	electronic warfare support
EW	electronic warfare

F

FBI	Federal Bureau of Investigation
FEBA	forward edge of the battle area
FIS	foreign intelligence service
FLOT	forward line of own troops
FM	field manual; frequency modulation
FORSCOM	United States Army Forces Command
FSC	foreign SIGINT collector

G

G2	Assistant Chief of Staff, G2 (Intelligence)

G3	Assistant Chief of Staff, G3 (Operations and Plans)
G4	Assistant Chief of Staff, G4 (Logistics)
G5	Assistant Chief of Staff, G5 (Civil Affairs)
govt	government
GS	general support

H

HA	humanitarian assistance
HHOC	headquarters, headquarters and operations company
HPT	high-payoff target
HQ	headquarters
HQDA	Headquarters, Department of the Army
HUMINT	human intelligence
HVT	high-value target

I

I&S	intelligence and surveillance
I&W	indications and warning
ICF	intelligence contingency funds
ID	identification
IEW	intelligence and electronic warfare
IFF	identification, friend or foe
IMFR	Investigative Memorandum for Record
IMINT	imagery intelligence
INCL	inconclusive
inf	infantry
INS	Immigration and Naturalization Service
INSCOM	United States Army Intelligence and Security Command
intel	intelligence
intg	interrogation
INTSUM	intelligence summary
IPB	intelligence preparation of the battlefield
IPW	prisoner of war interrogation
IR	information requirements
IRR	Investigative Records Repository

J

J	jamming

J2	Intelligence Directorate
JCS	Joint Chiefs of Staff
J-TENS	Joint Tactical Exploitation of National Systems
JTFCICA	Joint Task Force CI Coordinating Authority

K

KIA	killed in action

L

LAA	limited access authorization
LEA	law enforcement agency
LIC	low-intensity conflict
LLSO	low-level source operation
LNO	liaison officer
LOS	line of sight
LZ	landing zone

M

M meter	
MC	multichannel
MBA	main battle area
MDCI	multidiscipline counterintelligence
MDCISUM	MDCI summary
METT-T	mission, enemy, troops, terrain and weather, and time available
MI	military intelligence
MIJI	meaconing, intrusion, jamming, and interference
MO	modus operandi
MOS	military occupational specialty
MP	military police
MRE	meals ready to eat
MSE	mobile subscriber equipment
msg	message
MTI	moving target indicator
MTOE	modification table of organization and equipment

N

NAC	national agency check
NAI	named areas of interest
NATO	North Atlantic Treaty
NBC	nuclear, biological, and chemical
NCIS	Naval Criminal Investigation Service
NCO	noncommissioned officer
NDI	no deception indicated
NGIC	National Ground Intelligence
NKSOF	North Korean Special Operations Forces
NO	no opinion
no	number
NRT	near-real time

O

OB	order of battle
OCONUS	outside continental United States
OJE	on-the-job experience
OOTW	operations other than war ops operations
OPLAN	operations plan
OPORD	operations order
OPSEC	operations security
OSI	Office of Special Investigations

P

PCS	permanent change of station
PFC	private first class
PIR	priority intelligence requirements
PL	phase line
plt	platoon
PSG	platoon sergeant
PSI	personnel security investigation
PSYOP	psychological operations

Q

qty	quantity

R

RAF	Royal Air Force
RC	Reserve Components
RDTE	research, development, test, and evaluation
RECCE	reconnaissance
REC	radio electronic combat
RF	radio frequency
RII	request for intelligence information
RISTA	reconnaissance, intelligence, surveillance, and target acquisition
rqr	requirement

S

S2	Intelligence Officer (US Army)
S5	Civil Affairs Officer (US Army)
SA	special agent
SAEDA	Subversion and Espionage Directed Against US Army and Deliberate Security Violations
SALUTE	size, activity, location, unit, time, and equipment
SAP	special access program
SATRAN	see FM 34-5 (S) for classified identification
SCA	special category absentees
SCARF	standard collection asset request format
SCI	sensitive compartmented information
SCO	sub-control office
sec	section
sgt	sergeant
SIGINT	signals intelligence
SIGSEC	signals security
SJA	Staff Judge Advocate
SLAR	side looking airborne radar
SM	service member
SMU	special mission unit
SOFA	Status of Forces Agreement
SOP	standing operating procedure spec specialist
SSB	single sideband
SSBI	single scope background investigation
SSN	social security number

STANAG   Standardization Agreement svc service

T

tac	tactical
TAO	tactical agent operation
TDA	Tables of Distribution and Allowances
TDY	temporary duty
TEB	tactical exploitation battalion
TOC	tactical operations center
TOE	tables of organization and equipment
TRRIP	Theater Rapid Response Intelligence Package
TSCM	technical surveillance countermeasures
TTP	tactics, techniques, and procedures
TV	television

U

UAV	unmanned aerial vehicle
UCMJ	Uniform Code of Military Justice
US	United States (of America)
USAF	United States Air Force
USAI	United States Army Intelligence
USAIC&FH	US Army Intelligence Center and Fort Huachuca
USAR	United States Army Reserve
USMTF	US message text format

V

VCR   video cassette recorder

W

wpn   weapon

X

xplat   exploitation

Section II. Terms

Analysis—A stage in the intelligence cycle in which information is subjected to review in order to identify significant facts and derive conclusions therefrom.

Assessment—Analysis of the security, effectiveness, and potential of an existing or planned intelligence activity.

Collection Intelligence Cycle—Acquisition of information and the provision of the information to processing or production elements.

Communications Intelligence (COMINT)—Technical and intelligence information derived from foreign communications by other than the intended recipients.

Communications Security (COMSEC)—The protection resulting from all measures designed to deny unauthorized persons information of value which might be derived from the possession and study of telecommunications, or to mislead unauthorized persons in their interpretation of the results of such possession and study. Communications security includes— cryptosecurity; transmission security; emission security; and physical security of communications security materials and information.

Compromising Emanations—Unintentional intelligence-bearing signals which, if intercepted and analyzed, disclose national security information transmitted, received, handled, or otherwise processed by an information-processing system.

Counterintelligence (CI)—Information gathered and activities conducted to protect against espionage, other intelligence activities, sabotage or assassinations conducted for or on behalf of foreign powers, organizations or persons, or international terrorist activities, but not including personnel, physical, document or communications security programs. Synonymous with Foreign Counterintelligence.

Counterintelligence (DOD, NATO) (JCS Pub 1)—
Those activities which are concerned with identifying and

278

counteracting the threat to security posed by hostile intelligence services or organizations, or by individuals engaged in espionage, sabotage, subversion, or terrorism.

Counterintelligence (Inter-American Defense Board) (JCS Pub 1)—That phase of intelligence covering all activity devoted to destroying the effectiveness of inimical foreign intelligence activities and to the protection of information against espionage, personnel against subversion, and installations or material against sabotage.

CI Liaison—The establishment and maintenance of personal contacts between CI liaison officers and personnel of organizations which have missions, responsibilities, information resources, or capabilities similar to those of US Army intelligence. It is conducted to promote cooperation, unity of purpose, and mutual understanding; coordinate actions and activities; and to exchange information and viewpoints. OCONUS CI liaison also includes overt collection of foreign intelligence and CI; acquisition from foreign sources of material and assistance not otherwise available; and the procedures used to gain access to individuals whose cooperations, assistance, or knowledge are desired.

Countermeasures—That form of military science that by the employment of devices or techniques, has as its objective the impairment of the operational effectiveness of enemy activity.

Counter-Signals Intelligence (C-SIGINT)—Those actions taken to determine enemy SIGINT capabilities and activities, the assessment of friendly operations to identify patterns and signatures, and the resulting vulnerabilities for subsequent development and recommendation of countermeasures. Recommendations to counter the foreign SIGINT collector (FSC) and EW threat are provided to the G3 by the G2. They can include offensive measures such as electronic attack, to include jamming or deception; or targeting for fire or maneuver.

Critical Node—An element, position, or communications entity whose disruption or destruction immediately degrades the ability of a force to command, control, or effectively conduct combat operations.

Cryptosecurity—The component of COMSEC which results from the provision of technically sound cryptosystems and their proper use.

Deception—Those measures designed to mislead the enemy by manipulation, distortion, or falsification of evidence to induce them to react in a manner prejudicial to their interests.

Doctrine—Fundamental principles by which the military forces or elements thereof guide their actions in support of national objectives. It is authoritative but requires judgment in application.

Electronic Protection (EP)—That division of electronic warfare involving actions taken to protect personnel, facilities, and equipment from any effects of friendly or enemy employment of electronic warfare that degrade, neutralize, or destroy friendly combat capability. Formerly known as electronic counter-countermeasures (ECCM).

Electronic Attack (EA)—That division of electronic warfare involving the use of the electromagnetic or directed energy to attack personnel, facilities, or equipment with the intent of degrading, neutralizing, or destroying enemy combat capability. Formerly known as electronic countermeasures (ECM).

Electronic Deception—The deliberate radiation, reradiation, alteration, suppression, absorption, denial, enhancement, or reflection of electromagnetic energy in a manner intended to convey misleading information and to deny valid information to an enemy or to enemy electronics-dependent weapons. Among the types of electronic deception are: manipulative electronic deception, simulated electronic deception, and imitative deception.

Electronic Jamming—The deliberate radiation, reradiation, or reflection of electromagnetic energy for the purpose of disrupting enemy use of electronic devices, equipment or systems.

Electronic Security (ELSEC)—The protection resulting from all measures designed to deny unauthorized persons information of value that might be derived from their interception and study of noncommunications electromagnetic radiations, for example, radar.

Electronics Intelligence (ELINT)—Technical and intelligence information derived from foreign noncommunications electromagnetic radiations emanating from other than nuclear detonations or radioactive sources.

Electronic Warfare (EW)—Military action involving the use of electromagnetic energy to determine, exploit, reduce, or prevent hostile use of the electromagnetic spectrum and action which retains friendly use of electromagnetic spectrum.

Emission Control—The selective and controlled use of electromagnetic, acoustic, or other emitters to optimize C 2 capabilities while minimizing, for OPSEC, detection by enemy sensors; to minimize mutual interference among friendly systems; or to execute a military deception plan.

Emission Security—That component of COMSEC which results from all measures taken to deny unauthorized persons information of value which might be derived from intercept and analysis of compromising emanations from crytoequipment and telecommunications systems.

Essential Elements of Friendly Information (EEFI)—Key questions about friendly intentions and military capabilities likely to be asked by opposing planners and decisionmakers in competitive circumstances.

Foreign SIGINT Collector (FSC)/EW—A foreign entity employing electromagnetic and SIGINT techniques to target friendly forces for the purposes of detecting, exploiting, or subverting the C-E environment of the friendly commander.

f-stop—A camera lens aperture setting indicated by an f-number.

Human Intelligence (HUMINT)—A category of intelligence information derived from human sources.

Imagery Intelligence (IMINT)—The collected products of imagery interpretation processed for intelligence purposes. Indicator - (1) In intelligence usage, an item of information that reflects the intention or capability of a potential enemy to adopt or reject a COA. (2) Activities that can contribute to the determination of a friendly COA.

Intelligence—The product resulting from the collection, processing, integration, analysis, evaluation and interpretation of available information concerning foreign countries or areas.

Liaison—That contact or intercommunication maintained between elements of military forces to ensure mutual understanding and unity of purpose and action.

Liaison Contact—The act of visiting or otherwise contacting a liaison source.

Liaison Officer—A CI special agent (SA) assigned the mission of conducting CI liaison.

Liaison Source—An individual with whom liaison is conducted. This term applies regardless of whether the individual furnishes assistance or is contacted on a protocol basis.

Mission—(1) The task, together with the purpose, that clearly indicates the action to be taken and the reason therefore. (2) In common usage, especially when applied to lower military units, a duty assigned to an individual or unit to task.

Operations Security (OPSEC)—The process of denying adversaries information about friendly capabilities and intentions by identifying, controlling, and protecting indicators associated with planning and conducting military operations and other activities.

Patterns—Stereotyped actions which so habitually occur in a given set of circumstances that they cue an observer, well in advance, to either the type of military unit or activity, its identity, capabilities or intent. Stereotyping occurs in a variety of ways, such as communications deployment techniques or historical association. Patterns must be unique and detectable to be of military significance.

Profile—The picture formed through the identification and analysis of elements, actions, equipment, and details of military units or activity. Pattern plus signature equals profile.

Risk—A measure of the extent to which a recommended countermeasure has been historically effective in eliminating a vulnerability, given a certain level of susceptibility and threat.

Security—(1) Measures taken by a military unit, an activity or installation to protect itself against all acts designed to, or that may, impair its effectiveness. (2) A condition that results from the establishment and maintenance of protective measures that ensure a state of inviolability from hostile acts or influences. (3) With respect to classified matter, it is the condition that prevents unauthorized persons from having access to official information that is safeguarded in the interests of national security.

Signals Intelligence (SIGINT)—A category of intelligence information comprising all communications intelligence, electronic intelligence, and telemetry intelligence.

Signals Security (SIGSEC)—A generic term that includes both COMSEC and ELSEC.

Signature—The identification of a military unit or activity resulting from the unique and detectable visual, imagery, electromagnetic, olfactory, or acoustical display of key equipment normally associated with that type unit or activity.

source—(1) A point of origin or procurement. (2) One that initiates.

Source—Any person who furnishes intelligence information either with or without the knowledge that the information is being used for intelligence purposes. In this context, a controlled source is in the employment or under the control of the intelligence activity and knows that the information is to be used for intelligence purposes.

SUBJECT—(1) A person who is the principal object of attention. (2) One who is under investigation.

Susceptibility—The degree to which a device, equipment, or weapons systems is open to effective attack due to one or more inherent weaknesses.

Threat—The technical and operational capability of a FSC or EW system to detect, exploit or subvert friendly signals and the demonstrated, presumed or inferred intent of that system to conduct such activity.

Vulnerability—Characteristics of a friendly C-E system or cryptosystem which are potentially exploitable by FSC or EW systems. As applied in this manual, vulnerability is a susceptibility in the presence of a threat. Susceptibility in the absence of a threat does not constitute a vulnerability.

References

Sources Used

These are the sources quoted or paraphrased in this publication.

Joint Publications

1-02. DOD *Dictionary of Military and Associated Terms.*
1 December 1989.
2.0. *Joint Doctrine for Intelligence Support to Operations.*
(S)2-01.2. Joint TTP for CI Support to Operations (U).

Army Publications

AR 1-100. *Gifts and Donations.* 15 November 1983
AR 15-6. *Procedures for Investigating Officers and Boards of Officers.* 11 May 1988.
AR 25-50. *Preparing and Managing Correspondence.*
21 November 1988.
AR 27-1. *Judge Advocate Legal Service*
AR (C)105-2. *Electronic Counter-countermeasures (ECCM) - Electronic Warfare Susceptibility and Vulnerability (U).*
30 September 1976.
AR 190-6. *Obtaining Information from Financial Institutions.*
15 January 1982.
AR 190-13. *The Army Physical Security Program.*
30 September 1993.
AR 190-22. *Searches, Seizures, and Disposition of Property.*
1 January 1983.
AR 190-53. *Interception of Wire and Oral Communications for Law Enforcement Purposes.* 1 November 1978.
AR 195-5. *Evidence Procedures.* 28 August 1992.
AR 195-6. *Department of the Army Polygraph Activities.*
1 September 1980.
AR 310-50. *Authorized Abbreviations and Brevity Codes.*
15 November 1985.
AR 380-5. *Department of the Army Information Security Program.* 25 February 1988.
(O)AR 380-40. *Policy for Safeguarding and Controlling COMSEC Material (U).* 29 July 1994.

AR 380-53. *Communications Security Monitoring.* 15 November 1984.

AR 380-67. *The Department of the Army Personnel Security Program.* 9 September 1988.

AR 381-10. *US Army Intelligence Activities.* 1 July 1984.

AR 381-12. *Subversion and Espionage Directed Against The US Army (SAEDA).* 15 January 1993.

(S)AR 381-14. *Technical Surveillance Countermeasures (TSCM) (U).* 3 October 1986.

AR 381-20. *US Army Counterintelligence Activities.* 26 September 1986.

(S)AR 381-47. *US Army Offensive Counterespionage Activities (U).* 30 July 1990.

(C)AR 381-141. *Intelligence Contingency Funds (U).* 30 July 1990.

AR 525-13. *The Army Combatting Terrorism Program.* 26 June 1992.

FM 6-20-10. *Tactics, Techniques, and Procedures for the Targeting Process.* 29 March 1990.

FM 11-65. *High Frequency Radio Communications.* 31 October 1978.

FM 19-20. *Law Enforcement Investigations.* 25 November 1985.

FM 19-30. *Physical Security.* 1 March 1979.

FM 24-16. *Communications-Electronics Operations, Orders, Records and Reports.* 7 April 1978.

FM 24-17. *Tactical Records Traffic System (TRTS).* 17 September 1991.

FM 24-18. *Tactical Single-Channel Radio Communications Techniques.* 30 September 1987.

FM 24-33. *Communications Techniques: Electronic Counter-Countermeasures.* 17 July 1990.

FM 34-1. *Intelligence and Electronic Warfare Operations.* 27 September 1994.

FM 34-2. *Collection Management and Synchronization Planning.* 8 March 1994.

FM 34-3. *Intelligence Analysis.* 15 March 1990.

(S)FM 34-5. *Human Intelligence and Related Counterintelligence Operations (U).* 29 July 1994.

FM 34-10. *Division Intelligence and Electronic Warfare Operations.* 25 November 1986.

FM 34-25. *Corps Intelligence and Electronic Warfare Operations.* 30 September 1987.

FM 34-25-3. *All-Source Analysis System (ASAS).* (TBP, June 1995.)

FM 34-35. *Armored Cavalry Regiment (ACR) and Separate Brigade Intelligence and Electronic Warfare (IEW) Operations.* 12 December 1990.

FM 34-37. *Echelons Above Corps (EAC) Intelligence and Electronic Warfare (IEW) Operations.* 15 January 1991.

FM 34-52. *Intelligence Interrogation.* 28 September 1992.

FM 34-130. *Intelligence Preparation of the Battlefield.* 8 July 1994.

FM 100-5. *Operations.* June 1993.

FM 100-20. *Military Operations in Low Intensity Conflict.* 5 December 1990.

FM 101-5. *Staff Organization and Operations.* 25 May 1984.

DA Pam 25-30. *Consolidated Index of Army Publications and Blank Forms.* 1 October 1994.

TC 34-55. *Imagery Intelligence.* 3 October 1988.

(S)TB 380-6-1-4. *Signal Security (SIGSEC) Advisors Package- SIGSEC Technical Support (U).* 14 January 1985.

Miscellaneous Publications

Constitution of the United States.

Executive Order 10450, Security Requirements for Government Employees.

Executive Order 12333, US Intelligence Activities.

Federal Sedition Statute.

Geneva Conventions.

Joint Tactical Exploitation of National Systems (J-TENS) Manual.

United States Code.

Uniform Code of Military Justice (UCMJ).

United States Manual for Courts-Martial.

United States Message Text Format (USMTF) User's Handbook.

NACSI4000. Military Rule of Evidence (315(d).

Privacy Act of 1974.

(U) DIAM 58-13. *DOD HUMINT Management System, Secret/NOFORN.*

DOD Reg 5200.1R. *DOD Information Security Program.* June 1986.

DODD 4640.6. *Communications Security Telephone Monitoring and Recording.*

DODD 5000.1. *Defense Acquisition.*

DODD 5200.27. *Acquisition of Information Concerning Persons Organizations Not Affiliated with The Department of Defense.*

DODD 5240.5. *Department of Defense Technical Surveillance Countermeasures Survey Program.*

Miscellaneous Forms

DA 2802. *Polygraph Examination Report.*
DA 2823. *Sworn Statement.*
DA 3881. *Rights Warning Procedure/Waiver Certificate.*
DD 173. *Joint Message Form.*
DD 398. *Personnel Security Questionnaire.*
DD 398-2. *Personal Security Questionnaire (National Agency Check).*

Standardization Agreements (STANAG)*

2033. *Interrogation of Prisoners of War.*
2044. *Procedures for Dealing with Prisoners of War.*
2067. *Control and Return of Stragglers.*
2079. *Rear Area Security and Rear Area Damage.*
2101. *Establishing Liaison.*
2363. *Security Doctrine.*
2844 *(Edition Two). Counterintelligence Procedures.*

STANAGs can be obtained from the Naval Publications and Forms Center (NPFC), 5801 Tabor Avenue, Philadelphia, PA 19120. Use DD Form 1425 to requisition documents.

Documents Needed

These documents must be available to the intended users of this publication.

DA 2028. *Recommended Changes to Publications and Blank Forms.*
(S)DIAM DJS-1400-7-85. *SATRAN (U).*
INSCOM Reg 381-6. *United States Army Intelligence and Security Command Polygraph Activities.*
DIS 20-1-M. *Manual for Personnel Security Investigations.*